Environment and Natural Resource Conservation and Management in Mozambique

Munyaradzi Mawere

Langaa Research & Publishing CIG
Mankon, Bamenda

Publisher:

Langaa RPCIG
Langaa Research & Publishing Common Initiative Group
P.O. Box 902 Mankon
Bamenda
North West Region
Cameroon
Langaagrp@gmail.com
www.langaa-rpcig.net

Distributed in and outside N. America by African Books Collective
orders@africanbookscollective.com
www.africanbookcollective.com

ISBN: 9956-790-77-X

Dedication

To Pedro, and my dearest cousin, John, I humbly and gratefully dedicate this book. Thanks for the inspiration

Contents

Acknowledgements

With the exception of the introduction, chapters 1, 5,6,7 and 8, the other chapters of this book appeared previously in the following journals, but of course with several modifications: 'A Critical Investigation of Environmental Malpractices in Mozambique: A Case Study of Xai-Xai Communal Area, Gaza Province, Mozambique', *International Research Journals,* Vol. 2 (2): 874-883 (Chapter 2); 'Green Revolution Program (GRP) In Mozambique: Rethinking the Impact of Mozambique's Fast-Track GRP on the Environment and Biodiversity', *Journal of Ecology and the Natural Environment,* Vol. 2 (6) 92-98 (chapter 3); and 'Gold Panning In Central Mozambique: A Critical Investigation of the Effects of Gold Panning In Manica Province', *International Journal of Politics and Good Gorvernace,* Vol. 2, No. 2.4 Quarter iv, 2011 (chapter 4). My thanks are due to the publishers for allowing me to reprint the chapters in this book, of course with some modifications in both their titles and contents.

I would like also to thank the reviewers for their meticulous reading of the first drafts of the book manuscript and for their constructive comments and suggestions.

Preface

The debates around issues of environmental and natural resource conservation and management in Africa and beyond have been highly momentous and have sustained controversies of epic proportions in conservation sciences and environmental anthropology. Given the nebulous nature of the concept of environment coupled with problems around resource conservation and management, a robust comprehension of both the concept of environment and root causes of the problems around natural resource conservation and management is more urgent than ever, especially in contexts such as those of emerging economies such as Africa in general and in particular Mozambique.

In Africa, the debate on environmental conservation has on one hand been provoked by the fact that while developing countries have contributed less than any other region to the greenhouse gas emissions that are widely held responsible for global warming, they are more vulnerable due to their dependence on burning fuels[1], and also because of their high poverty levels. As I pointed out elsewhere[2], 'the impacts of climate change in Africa are generally manifested in human health and in the agricultural sector worsening the existing levels of poverty and undermining all development efforts in the continent'. The other consequences include adverse impacts on Africa's varied livelihoods systems and the unique biodiversity of its ecosystems. In fact the global environmental change such as climate change, for example,

---

[1] See Mawere, M. 2010. In a recent study on environmental malpractices in Mozambique first published in *International Research Journals,* Vol. 2 (2): 874-883.

[2] Ibid.

has impacted negatively on ecosystems thereby making water and food security especially in already semi-arid areas more uncertain than ever. This has caused most forecasting scenarios suggesting greater vulnerability to ecosystem damage, reduced ecosystem services, and undermined resilience to global environmental change especially among the rural poor. Building resilience to global environmental change and other shocks, for example, needs to be mainstreamed into the rural people's livelihood planning to ensure self sustenance and food security. This is especially important for vulnerable populations with low adaptive capacities such as many of the rural people in sub-Saharan African countries like Mozambique. The majorities of these people make their living from rain-fed agriculture[3], and largely depend on small-scale subsistence agriculture for their livelihood security[4].

On the other hand, debate on natural resource conservation and management has been chiefly sparked by the paradox that while there seem to be a 'natural resource boom in Africa', the continent remains one of the poorest in the world. One of Africa's countries, Mozambique for example, is ranked one of the poorest countries in the world. In fact, as the Rural Poverty Report-Mozambique (2007) and Africol.com[5] revealed, Mozambique remains one of the poorest nations in the world and more than 80% of its citizens in rural areas live in extreme poverty, living on less than a US$1 a day, and lack basic services like schools, clean

---

[3] See study by FAO. 19995. *World agriculture: Towards 2010*, p. 481.

[4] This is elaborated by Rockstrom in his article: Water resources management in smallholders farms in Eastern and Southern Africa, 25 (3): 275-283.

[5] For more details on rural poverty in Mozambique read a report by Afrol.com, 2008. The report is available at: http://www.afrol.com/Mozambique.

water and hospitals. Although the concept of extreme poverty is difficult to define with precision as it is not static, I identify with the World Bank's 2005 and 2010 definition that 'extreme poverty is living on less than US$1.25 a day. This means living on the equivalent of US$1.25 a day, in the US, buying US goods. In 2011, this meant surviving on the equivalent of US$1.50, AUD$2 or 1 pound per day'[6]. Thus, according to this definition, majority of Mozambique's population, as in many other parts of Africa, live in extreme poverty, yet Mozambique is one such country that is endowed with diverse natural resources including fisheries, forests, wildlife and minerals[7].

In view of these observations, this book generally explores environmental and natural resource conservation and management issues in developing countries in Africa. In particular, the book explores environmental and natural resource conservation and management problems in rural communities of Mozambique. While the problems discussed in this book are common phenomena in many countries in Africa and beyond, the book adopts Mozambique as its case study for two major reasons. First, the choice of Mozambique is premised on the fact that Mozambique's environment and conservation problems have persisted from her protracted colonial experience since around 1500 through independence to the present time. Second, as noted by Leslie[8], social

---

[6] For detailed information on extreme poverty, see World Bank. (2010). Extreme poverty rates continue to fall, Available online at: http://www.worldbank.org.

[7] The richness of Mozambique in terms of natural resources is highlighted in Mawere, M. 2010; in a recent study on environmental malpractices in Mozambique first published in *International Research Journals,* Vol. 2 (2): 874-883.

[8] For further discussion on social systems, stress and shocks on the environment see Leslie, 2008. 'Conservation in Flux: Pursuing Social

systems constantly shift to exert new stresses and shocks as well as present unanticipated opportunities that can impact environment, biodiversity, and natural resource conservation efforts. In view of this understanding, it could be argued that Mozambique being one country whose social systems have kept on changing since the colonial period through the present times conservation strategies need to be re-examined regularly to ensure that they are meeting their objectives and applicability in context. Besides, studying and understanding such long standing troubling issues as environmental and conservation problems of countries such as Mozambique, which stretch from the colonial period to the present, affords the opportunity for environmentalists and policy makers in countries/contexts with similar problems.

In the attempt to unpack the conservation status of Mozambique, the present book starts by situating the country's environment and natural resource conservation in its historical context. The book goes on to examine different environmental and natural resource conservation problems using selected Mozambican case studies, studied through diverse methodologies ranging from ethnography to surveys, field observations, interviews, informal discussions and documentary literature. From the respective findings, the book has as its running thread the thesis that environmental and natural resource conservation and management in Mozambique as in many other African countries that experienced colonialism has been in a state of crisis since the colonial period. This is not to say that the book is uncritical of the present environmental and resource conservation and management systems as well as the associated issues. In fact, the book also notes that the present environment and

Resilience in Mozambique and Peru', the Bulletin of the Yale Tropical Resource Institute, Volume 27.

resource conservation crises are not only a result of the colonial legacy, but other many factors.

The reasons for the extension and persistence of the crisis through time are identified using various case studies used in the essays that make part of this book. These range from poor governance, global environmental changes, poverty, corruption by elites in areas of conservation, poor resource base – both monetary and human resource with conservation expertise, lack of government commitment to institute reforms and enforce them fully, among others. In terms of poor governance and corruption, for example, it is noted that apart from battling with business and business ethics cleavages of her 'long suffering' economy, Mozambique faces excruciating demands of political and socio-economic democracy which it is failing to meet as a country mainly as a result of external forces, poor governance and corruption by some government officials. This ascends the pedestal and gravity of the country's environment and resource conservation crises. The grand question, however, is whether we will, in the future, ever talk about sustainable environmental and natural resource conservation in Mozambique or at least conservation in any meaningful and coherent manner if poor governance and corruption persist in areas of environment and resource conservation as is in other such areas.

In addressing this question, I have assumed that many countries in the African continent that were affected by colonialism are bound to have more environmental and resource conservation problems in common than with countries of other continents whose environments and resources have never suffered the adverse impacts of colonialism. Also, I have not pretended to say everything about environmental and natural resource conservation in societies that directly experienced colonialism such as Mozambique except that which directly intertwine with the theme of this book. I have used some references and/or examples of certain phenomena common in some countries other than drawn from Mozambique so as to elucidate my arguments, but making sure not to lose focus in my treatment of environmental and natural resource conservation in Mozambique.

The book is a valuable asset for environmental conservationists, land resource managers, social ecologists, environmentalists, environmental anthropologists, environmental field workers and technicians, and practitioners and students of conservation sciences. It shows them the wisdom that lies in 'critical thinking' around environment related issues and encourages them to start on the path of critical thinking and careful reflection on issues that affect the environments and resource management in their respective contexts. The book, above all else, attempts to show that questioning, and indeed right questioning, matters most, across disciplines including conservation sciences and environmental anthropology.

Guide to the Reader

Environment and Natural Resources Conservation and Management in Mozambique is based on material that deals with issues of environmental and natural resource conservation and management in Mozambique. Using case studies and researches based on ethnographic studies, observations, interviews, informal discussions and documentary literature, the book advances the thesis that environmental and natural resource conservation crises in Mozambique have a long sad history that goes back to the colonial era. Yet while advancing this thesis, the book is quick to criticize the thinking by some environmentalists that ecosystem destruction, land degradation and natural resource over-exploitation are solely a result of inadequate understanding of the true value of resources by the rural poor and other such users. In light of this understanding, the book argues that deforestation, environmental degradation and overexploitation of some natural resources is a result of many factors (including limited knowledge of resource value), most of which are discussed in this book. While urging other scholars to search for more lasting but local solutions to the problems associated with environmental and natural resource management in the country, the book suggests that government commitment and local community participation in natural resource management issues are imperative and effective in arresting the conservation crises that Mozambique is facing. These recommendations are emphasized directly or otherwise in all the chapters that make part of this book. The book presents eight chapters structured as follows:

Chapter one examines the history of natural resource conservation in Mozambique since the colonial era. This historical overview is significantly important in so far as "local' natural resource conservation and environmental management cannot be understood without taking a historical approach, looking at social relations, and looking beyond the local system to wider political-economic structures" [9]. To this end, I assert that environmental and natural resource conservation crises that Mozambique experienced during the colonial era did not end with the demise of colonial administration at the country's independence in 1975. It has become a legacy carried into post-independence period, particularly the civil war (1976-1992) and post-civil war periods (1992-to date). In fact neither the conservation of natural resources nor management of the environment by respective authorities in post-civil war periods has significantly improved, as evidenced by the ever-increasing deforestation, illegal logging, land degradation, illegal mining, illegal hunting/poaching and overfishing currently taking place across the country. As such, the environmental and resource conservation trails that the Portuguese colonialists instigated some centuries ago are still observable in Mozambique even to date. As discussed in this book, the reasons for the extension and persistence of these crises are not only complex but many ranging from external forces, poor governance, corruption by elites in areas of conservation, poor resource base – both monetary and human resource with conservation expertise –, lack of government 'political will' to institute reforms and enforce them fully, among others.

---

[9] See Dove, R. M., Sajise, E. P., and Doolittle, A. A. 2011. *Beyond the sacred forest-Complicating conservation in southeastern Asia*, Duke University Press, London, p. 24.

Chapter two investigates a plethora of environmental malpractices in Mozambique. While the chapter focuses on the environmental malpractices in Mozambique in general, particular reference is made to a rural area in the southern region of the country, Xai-Xai, in Gaza province. The main argument that runs throughout this chapter is that environmental mismanagement in Xai-Xai communal area mirrors environmental problems and natural resource mismanagement at national level. Cases of resource mismanagement are drawn from different sectors of the economy ranging from fishery, agriculture/land use, forestry and wildlife with the recommendation that government's full commitment is required to ensure effective and efficient co-management of the environment and resources in the country.

In chapter three, the question of the environment and social relations as well as interactions with and within the environment is discussed. It should be underscored that the concept of environment as is problematized in this book is a nebulous one that warrants a robust understanding and clarifying before looking at the discussion in chapter 3.

Though an oft-talked about, the concept of 'environment' has been defined differently across disciplines. In common usage, the concept is often used as a synonym for nature (i.e. the biophysical or nonhuman environment), but this usage creates great conceptual confusion because the environment of a particular human group includes both cultural and biophysical elements[10]. Taking into account this observation, this study uses the term environment as often used in environmental anthropology to refer to "an explicit, active

---

[10] For controversies associated with the concept of environment, see Rappaport, R.A. 1979. *Ecology, meaning, and religion,* Berkeley, CA: North Atlantic.

concern with the relationship between human groups and their respective cultural and biophysical elements"[11]. The main argument advanced in chapter 3 where issues relating to the environment as it relates to humans and other beings (living or nonliving) are discussed is that in the context of the current green revolution program in Mozambique – whose noble motive was to cushion food insecurity in the country – there is need to rethink the ecological problems prompted by the program. This argument is raised against the background that there is ample evidence of lack of consideration for wildlife, forests and/or the environment's 'rights' in general by most rural farmers involved in the current green revolution program in the country. Basing on the evidence available on how forests are destroyed and wildlife and other such other beings abused in the face of green revolution program, the chapter urges the government, through the responsible ministry, to rethink the revolution program which is described in this book as the 'fast-track green revolution program'. The name 'fast-track green revolution program' reveals the hastiness and lack of preparedness or proper planning associated with the programme.

Chapter four is a critique of the exploitation and conservation of a highly valued resource, gold in the central provinces of Mozambique, and in particular Manica. In this region, gold panning is a source of livelihood for many yet a government source of revenue as well. Results obtained during this study, however, reveal that despite the fact that gold panning is a source of livelihood for many people in central Mozambique and beyond, it has not succeeded in

---

[11] For the definition of environment adopted in this book, see Little, P. 1999. Environments and Environmentalisms in Anthropological Research: facing the new millennium. *Annual Review of Anthropology*, 29: 253-284.

many respects in improving the people's lives. Instead, gold panning has caused (and is still causing) serious life threatening problems to livestock, the locals and others who depend on river/water resources. The problems that are resulting from gold panning in central Mozambique range from water pollution, poisoning (with mercury) of aquatic creatures, increase in sexually transmitted diseases incidences through prostitution, and land degradation. Like the problems that arise from the green revolution program discussed in chapter three, the author recommends total commitment by the government to re-organize mining activities in this gold rich region of the country so as to at least minimize or totally do away with the problems emanating from the activity.

Chapter five focuses on community-based natural resource management (CBNRM) in Mozambique. While CBNRM in Mozambique is not as developed as in many other countries in the region especially in terms of institutional organization, projects scattered in central and northern provinces such as Community Management Projects of Ancuabe (Cabo – Delgado), Chipange Chetu (Niassa), and Sanhote (Nampula), and Pindanyanga (Manica) are given as exemplary cases where CBNRM principles are to some extent being successfully implemented. Yet these CBNRM projects are not typically organic but externally initiated by some development agents. Using these various examples and other such isolated cases in the country, the chapter advances the thesis that the government should encourage local community participation in issues of environmental and natural resource management if conservation crises in the country are to be overcome. Based on the data gathered during research for the present book, this chapter argues with

Folke et al [12], that local stakeholder participation, robust cross-scale linkages of institutions and organizations as well as effective leadership can contribute to environmental and resource conservation success. Thus as long as local community members feel like they are discriminated or excluded from participating in conservation projects in their own areas, such projects can never be successful.

In chapter six, the book takes up the issue of conservation and preservation of natural resources with a view to determine a better strategy for locals in rural resource constrained areas of Mozambique. The chapter takes up an example of dambos in the southern region of the country, Chokwe, where these resources (dambos) are being underutilized instead of exploited to the full to better the locals' lives, majority of whom are living below poverty datum line. As the title of the chapter suggests, the chapter advances the argument that in the Chokwe rural area there is a conservation crisis premised on the fact that neither preservation nor sustainable conservation of dambos is being practiced. In view of this observation, the chapter is concluded by urging the government and concerned stakeholders to educate people in the concerned rural areas and take them a step further to engage in sustainable conservation such that they protect the dambos while at the same time earn a living out of them.

Chapter seven examines the value of natural resources in sustaining socio-economic lives of the rural people. More importantly, the chapter, using data generated from surveys and field observations suggests that majority of the people in

---

[12] See Folke, C., T. Hahn, P. Olsson and J. Norberg. 2005. Adaptive governance of social-ecological systems. *Annual Review of Environment and Resources* 30: 441-473.

central Mozambique rely on agriculture, forestry, wildlife, water resources, and other such resources that occur naturally on land for livelihood. Yet due to global environmental changes, competition, poverty, external 'forces' and ignorance, these resources are often stressed. The underlying thesis of the chapter is, therefore, that the government and concerned stakeholders and/or parties should embark in a national resource management and environmental awareness campaign, especially in rural areas to raise awareness among rural populations on how such resources should be utilized for their own benefit. Such awareness campaigns will also help educate rural communities on the dangers of activities such as gold panning, uncontrolled hunting, deforestation and others, both to their own livelihoods and to the natural environment in general.

Finally, chapter eight suggests general recommendations but based on discussions in the different chapters on how environment and natural resource conservation in Mozambique can be improved. The chapter, which is the conclusion to this book as a whole, is more of a summary of the main points that have been discussed throughout the book. More importantly, the chapter points to the possible direction that Mozambique's future conservation project(s) or culture could take to ensure a healthy environment and success in human socio-economic development.

The book is important in three main ways, but not only limited to these: First, it attempts to address the rights, needs and interests of local people and more marginal groups in the Mozambican rural areas. Second, it makes a clarion call for participation of the local or marginal groups in conservation issues as one way to ensure that fundamental conservation objectives and sustainable conservation are achieved both in theory and practice. Third, the book develops guidelines for a more integrative and socially interactive approach to natural resource conservation and to environmental planning, management and implementation.

Chapter 1

Situating Mozambique's Environment and Natural Resource Management in Historical Context

It is generally agreed that before colonialism in Africa, the 'local' or indigenous people lived at peace with other beings (fauna and flora) in the natural environment they shared. Besides the fact that their populations were still small and their level of industrialization too low to negatively and significantly impact their environment, they used their natural resources sustainably following the rules or canons agreed upon by the whole group. This, however, should not be taken to mean there were no trespassers who sometimes break away from the group's canons and act in a way that threatened and put the natural environment at risk. Veld fires, accidentally caused or otherwise, were cases in point. Those who violated the canons were, however, either punished by traditional authorities or directly suffered the consequences from the wrath of angry spirits who were believed to cause death or spell out misfortunes on the perpetrator. Basing on such cosmologies, humans knew they were unique specie but interacted and related with other species, plants and animals, in a way that showed respect and symbiotic interdependence. This way of relational interaction with nonhuman entities however changed with the advent of colonialism in Africa. As Dzingirai and Breen noted, 'from colonization onwards, indigenous people [in the African continent] began relating to wildlife in a predatory way'[1]. This thinking and observation

---

[1] Quoted in Dzingirai, V. and Breen, C. 2004. 'The community-based natural resource management crisis and research agenda', pp. 1.

1

has been shared by many scholars such as Moyana[2], Moyo[3], and Child[4] who argue that Africans behaved differently during colonialism simply because the land for protected areas and white agriculture was taken from them, often with coercion, denying them (the rightful owners of the land) access to natural resources which naturally belonged to them. For Dzingirai and Breen, 'it is likely that the indigenous people's longing to eliminate protected resources and to populate the landscape with land-use practices benefiting the household rather than government, intensified in response to the dramatized injustice'[5] by the colonial regime in Africa. In many African countries, including Mozambique, the dramatized injustices especially the inequalities between races, particularly the indigenous Africans and Portuguese settlers resulted in the indigenous or local people changing their attitudes towards natural resources and their surroundings – environment. The philosophy and/or attitude that resulted from the inequalities and injustices instigated by the white settlers were that of resisting efforts by conservation organizations even after independence as the larger part of the contents of the colonial governments' constitutions and traits were simply inherited by most African governments after independence. This was simply because the attainment of independence by many African countries did not result in the automatic regain of their lost land and the re-writing of the national constitution. In many African countries, the black majority continued cultivating the infertile soils while the white minority who remained after independence continued using vast fertile lands they had gained during

---

[2] For further discussion on political economy see Moyana, H. 1984 in his *Political economy of land in Zimbabwe*.

[3] Moyo, S. 1995. *The land question in Zimbabwe,* Sapes Books: Harare.

[4] See Child, G. 1995. Wildlife and people: The Zimbabwean success - How conflict between animals and people became progress for both, Wisdom Institute: Harare.

[5] Quoted in Dzingirai, V. and Breen, C. 2004. 'The community-based natural resource management crisis and research agenda', pp. 2.

colonialism. In South Africa, for example, the Institute for Poverty, Land and Agrarian Studies (PLAAS) (2013) revealed that:

> *In 1994, as a result of colonial dispossession and apartheid, 87 % of the land was owned by whites and only 13 % by blacks. By 2012 post-apartheid land reform had transferred 7.95 million hectares into black ownership, which is equivalent, at best, to 7.5 % of formerly white-owned land. Whites as a social category still own most of the country's land and redressing racial imbalances in land ownership is land reform's most urgent priority.*

To fight the 'new' philosophy of resistance to "foreign" conservation by the local people, many post-independence governments resorted to force. In Kenya and Zimbabwe, for example, the service of WWF was sought with Kenya acquiring from it a helicopter to track and shoot poachers of the Serengeti, but this failed leaving conservationists doubtful if they would ever be able to come up with a 'new' more effective conservation mechanism[6] than those that were failing.

From the foregoing, it is beyond question why in many African countries (including Mozambique), conservation crises that were created by colonial regimes persisted even some decades after independence and to date. As shall be seen in the ensuing sections of this chapter, Mozambique's conservation crisis is historical; it is a legacy from the colonial era that has been precipitated with abject poverty, lack of government commitment and poor governance, among other factors. This chapter makes an attempt to situate environmental and natural resource conservation and management in history. It traces the conservation mechanisms and legislations instituted to give force to the

---

[6] See Bonner, R. 1993. *At the hand of man*, Alfred Knopf: New York.

mechanisms since the Portuguese rule in Mozambique to the present times. Such a historical overview of environmental and natural resource conservation and management is given to determine the source and trend of the conservation crisis that Mozambique and other countries in the region are facing.

Research question

The basic research question that this chapter seeks to examine, and perhaps run through other chapters of this book is: 'What is the status of environmental and natural resource conservation in Mozambique and why is it like that?' This question cannot be fully addressed without situating Mozambique's natural resource and environmental management in historical context, that is, from the colonial period when organic CBNRM – CBNRM as practised by the indigenous people – was disrupted through the present time. The sections that follow, therefore, trace the history of natural resources [including forests, wildlife, minerals and fisheries] conservation in terms of the legislations used and how these were employed to enforce fully or partially the conservation of natural resources and the environment in the country.

Legislations passed for the management of the environment and natural resources during the colonial era

Mozambique was a Portuguese colony for close to five hundred years (1505-1975) but with much concentration in the southern part of the country and along the Indian Ocean. Although modern boundaries of Mozambique were formed in the 19th century with colonial administration largely limited to the southern part of the country, and with much of the centre and north of the country granted as concessions to two vast chartered companies, the whole of Mozambique remained a Portuguese colony since the late 15th century. As Vines[7] noted 'only between 1941 when the last of the company charters elapsed and 1974 was Mozambique governed as a single administrative unit with a national economy'. It is during this period, 1941-1974 that the new state commonly known as *Estado Novo* (New State) was formed in Mozambique.

During this colonial era in Mozambique as elsewhere in colonized African states, there was very limited democracy as the colonial government had highly centralized authority and bureaucratic administration ideologies. Socio-economically, there were large disparities between the Portuguese settlers/people and the so-called "natives" or local black Mozambicans, with the larger population located in the rural communities where soils are poor – a distinctive colonial legacy that have been carried up to the present not only in Mozambique but many other African countries. This partly explains why majority of the Africans have remained poor or even poorer to date. In relation to Mozambique, the country is one of the poorest nations in the world[8] and more than 80% of its citizens in rural areas live on less than a US$1 a day, and lack basic services like schools, clean water and hospitals.[9] These people lack basic necessities to exploit resources around them in a way that would improve their

---

[7] For further discussion on the history of Mozambique see Vines, 1996. Cited in Anstey, S.G. 2004. 'Governance, natural resources and complex adaptive systems: A CBNRM study of communities and resources in northern Mozambique.

[8] More is available at Afrol.com website: *http://www.afrol.com/Mozambique*.

[9] See more in the Rural Poverty Report on Mozambique, 2007.

lives while not straining the resources and the environment in general.

While gold mining was the dominant economic feature in the 15th and 16th centuries, ivory trade dominated the Mozambican economy during the 17th and 18th centuries. This shift from gold to ivory was largely influenced by the fact that Mozambique was one of the countries rich in ivory, a premier resource of the time. Ivory promoted Indian, Arab and Portuguese influence along the coast and linkages into the interior.[10] Although the first wildlife legislation was passed for Portugal's colonies, including Mozambique in 1909, the legislation was largely limited to the licencing of hunting for certain commercially valued species, and not to restrict elephant hunting,[11] and hunting of other large and/or endangered species. Given that Portugal primarily depended much on wildlife as a source of meat for urbanites and farm/large-scale plantation labourers, and for exportation to sustain its economy, the colonial government attended the 1933 London Convention – the first international convention concerned with wildlife preservation – but did not ratify the Convention until 1949: it only issued a decree to implement its commitments in 1955[12]. As such, the 1955 Decree was established as the first wildlife legislation. The legislation declared that all wildlife was state property and could only be used under licence, and that the state should conserve wildlife and especially rare and endangered species by creating protected areas. However, the legislation was only ironical as the government issued commercial meat hunting licences to a

---

[10] See Newitt, M. 1995. *A history of Mozambique,* Hurst and Company, London.
[11] See Soto, B. 2003. Protected areas management in Mozambique, Report for IUCN-SASUSG, Harare.
[12] Ibid.

large number of Portuguese hunters with the result that during the 1950s there was what Spence described as 'uncontrollable destruction on such a scale that the country between the Zambezi and Ruvuma rivers, which used to team with game, has been almost completely denuded of wildlife'[13]. As further reported by Spence,[14] by the end of the 1950s the only remaining areas with significant wildlife were limited to the remoter parts of Niassa, south of Zambezia, the central provinces of Manica and Sofala, and inland sections of southern Mozambique.

This made the colonial government realize that the issuing of too many hunting licences to hunters had been mistakenly done. Having realized the negative effects of excessive hunting by the licenced hunters, the Portuguese government (in Mozambique) introduced new wildlife laws which at least gave power to the 1955 Decree in two major ways. First, the law reduced the number of large-scale commercial hunters to only 18 in the 1960s. As noted by Soto,[15] the government also identified 92, 200km^2 of protected areas comprising of national parks and sport hunting areas known as *Coutadas* (Conservancies) that were both reserved mainly for tourism. Second, wildlife management was transferred to the Veterinary Services Department in the country which besides having limited staff to control hunting in the whole of Mozambique was surprisingly responsible for wildlife elimination in the Save area[16]. This was a result of the hanging on to the mentality of

---

[13] See Spence, C.F. 1963. *Mozambique: East African province of Portugal*, Howard Tommins, Cape Town, pp. 16.
[14] Ibid.

[15] See Soto, B. 2003. Protected areas management in Mozambique, Report for IUCN-SASUSG, Harare.
[16] Opcite.

many hunters prior to the 1955 Decree. Besides identifying protected areas, the Portuguese government also strengthened the powers of its conservation legislations by integrating with some countries such as South Africa and the then Rhodesia (now Zimbabwe). This initiative brought to Mozambique new ideas on conservation of the environment and natural resources as well as management of revenues from tourism. Yet during all this time (the colonial era), only the Portuguese government and not the local black Africans were responsible for enacting and enforcing natural resources conservation legislative laws. This did not go well with the local black Mozambicans who felt they were denied the right to decision-making on issues to do with management of resources that were traditionally and legitimately theirs.

Environment and natural resource management after independence (1975-1992)

With the coming of independence in 1975, a lot of new changes were witnessed not only in the area of natural resource management, but in the economic and political systems of Mozambique. In terms of wildlife management, for example, by 1977, some of the inherited wildlife colonial legislations such as '*Legislação sobre as actividades da caça*' (Hunting activities legislation) were revised by the government in an attempt to create space for local community participation in natural resource management[17]. The revisions of this legislation emphasized the formation of local communities' cooperatives that can be granted low cost hunting permits instead of individuals (particularly the elites or politically well-connected) as the colonial government

---

[17] Opcite.

legislation dictated. The socialist system of cooperatives, thus, was introduced to replace the capitalistic system. The system of cooperatives was a socialist structure borrowed from socialist countries such as Russia and China which had greatly assisted Mozambique during its liberation struggle against the Portuguese colonialists. The *Coutadas* that had been formed during the Portuguese regime were also reorganized into Wildlife Utilization Units under a state enterprise called EMOFAUNA, responsible for sustainable production of wildlife meat for state farms, the military and urbanites[18]. It is worth noting, however, that this 'new' system was operational only for a few years before it was disrupted by the civil war in the early 1980s. The war was partly ignited by the socialist system that was introduced by the ruling party, *Frente de Libertação de Moçambique* (FRELIMO) government at independence in 1975 as some people viewed the system to be exploitive and cultivating poverty and laziness in the masses. As noted by Tello[19], 'state wildlife protected areas were reduced to covering 7 % of the country or around 56, 000km^{2}' as field staff were withdrawn from protected areas to the provincial capitals or the national capital, Maputo, during the height of the civil war. Some reports estimate that during the late 1970s to early 1990s, wildlife decreased by 90 % within the protected areas and in all the country's provinces,[20] especially in the central and northern provinces of the country.

---

[18] See Tello, 1996. Cited In Anstey, S.G. 2004. 'Governance, natural resources and complex adaptive systems: A CBNRM study of communities and resources in northern Mozambique'
[19] Ibid, pp. 162.
[20] Hatton, J., Couto, M., and Oglethorpe, J. 2001. Biodiversity and war: A case study of Mozambique, Biodiversity Programme, WWF, Washington DC.

During the same time, mining activities especially in the central provinces which were the strongholds of the opposition party, RENAMO were halted, but with other mineral resources such as gold looted out of the country as some national gold smiths and foreign investors in the mining sector saw no future for the country (see chapter on gold mining, this volume). Only at least forest products such as wood for carpentry and industrial use, which are heavy to transport, survived during this time as minerals, fisheries and wildlife resources were being looted out of the country by opportunists.

Environment and natural resources management trend in Mozambique since 1992

With the end of civil war in 1992, Mozambique's economy and other such institutions including its natural resources department had fallen to its knees. To restore sanity in the sector of environment and resource conservation management, the government paved way for a renewed and re-organized conservation system. The National Directorate of Forestry and Wildlife (DNFFB) with the aid of Food and Agricultural Program (FAO) and United Nations Development Program (UNDP), which had helped the country during civil war, sought to re-organize conservation of forestry and wildlife in the country by promoting sustainable exploitation of these important resources. However, by end of 1993, DNFFB had already been paralyzed by lack of funding from core donor projects[21] Conservation in the country thus faced another serious crisis that was partially ended in 1994 when IUCN sourced funds

---

[21]Soto, B. 2003. Protected areas management in Mozambique, Report for IUCN-SASUSG, Harare.

for the rehabilitation of protected areas such as Gorongoza National Park. The crisis was only partially solved because most of the country's protected areas were still understaffed and under-resourced with some falling short of money for workers' salaries and for general management. As reported by National Agricultural Program (PROAGRI, 1997), state funding for DNFFB declined from USD 1.4 million to 0.35 million between 1992 and 1995, 'and was scarcely sufficient to cover salaries, let alone management of natural resources or rehabilitation of protected areas'[22] As Anstey *et al*[23] noted, the period that followed saw the wider field of natural resources characterized by a flood of investors mainly from South Africa seeking and gaining licenses and concessions to natural resources that the state agencies were ill equipped to handle, thereby creating fertile grounds or conditions for lack of transparency and corruption by Mozambican elites. For most corrupt Mozambican elites, this was time for them to make money out of the country's natural resources.

The lack of resources, for example money for salaries, equipment etc., resulted in Mozambique privatizing some of its resources at provincial and national levels, and also encouraging community involvement in natural resource management. In terms of privatization, three programs that promised to inject a total of about US$31 million in natural resource management were formed. These were:

---

[22] For further discussion on natural resource governance see Anstey, S.G. 2004. 'Governance, natural resources and complex adaptive systems: A CBNRM study of communities and resources in northern Mozambique', In Dzingirai, V. and Breen, C. 2004. *Confronting the crisis in community conservation – Case studies from Southern Africa,* pp. 165.

[23] See a paper by Anstey, S. G., Abacar, J.A., and Chande, B. 2002. "It's all about power, it's all about money": Governance and resources in northern Mozambique.

1). *GERFA Projecto*

This project promised to inject about US$13.2 million to the forestry and wildlife resource management. The project was concerned with forestry and wildlife resources management focusing mainly on the rehabilitation of Gorongoza and Maromera protected/conservation areas, and the development of social forestry projects in central Mozambique.

2). *Support for Community Forestry and Wildlife Management Project*

This project was a joint venture between FAO and DNFFB, hence was widely known as FAO/DNFFB project. The project, which was funded by the Dutch government promised to inject about US$ 9.6 million. Its main objective was to establish a central community management unit training courses in Community-Based Natural Resource Management (CBNRM) in Universities in the country as well as running pilot CBNRM projects in Nampula and Maputo provinces. A broad definition of CBNRM as given by Murombedzi[24] asserts that CBNRM generally applies to 'a wide range of interventions that are designed to improve the management of natural resources in communal tenure regimes, through devolution of certain rights to these resources, and for the ostensible benefit of the owners and thus managers of these resources'. A more or less the same definition of CBNRM was given by Hviding and Jul-Larsen (1995) in their book, *Community resource management in tropical fisheries*, as initiatives by the state or development agents to accomplish resource management objectives through

---

[24] See Murombedzi, J.C. 2003. 'Pre-colonial and colonial conservation practices in southern Africa and their legacy today', pp. 12.

encouraging and facilitating the participation of rural communities. The above definitions of CBNRM have the implication that though the owners and managers of natural resources might be the local communities, conservation and management of the resources are externally initiated either by development agents or the state and not the local communities themselves. Yet prior to the establishment of CBNRM as externally initiated, CBNRM in many African societies was organic, that is, CBNRM as initiated and practised by the indigenous people who own and use the natural resources in their respective communities (see Katerere 1999; Marongwe 2004).

3). *Transfrontier Conservation Areas Project (TFCA)*

This project which was funded by the World Bank focused on protected areas in the southern region of the country. The project promised to inject about US 8.1 million. One of the major objectives of this project was to develop socio-economic networks with other conservation areas in the region,[25] especially in the neighbouring countries such as South Africa and Zimbabwe.

While the above projects focused mainly on natural resources such as forestry and wildlife, an integrated donor – Mozambique government investment program known as National Agricultural Program (PROAGRI) also existed focusing on agriculture and fisheries and also covering forestry and wildlife[26] which during this time all fell under one ministry – the Ministry of Agriculture and Fisheries. The major objective of PROAGRI was to ensure sustainability

---

[25] See also Anstey, S.G. 2004. 'Governance, natural resources and complex adaptive systems: A CBNRM study of communities and resources in northern Mozambique.
[26] PROAGRI, 1997. *The forestry and wildlife sector,* MINADER, Maputo.

and equality in resource management between all the three stakeholders involved (the three projects mentioned above) in resource utilization and environment management in the country. The initial phase of the program stretched from 1998-2002 and with the goal to achieve sustainable exploitation of natural resources and management of the natural environs in the country.

This goal was however not achieved as the Ministry responsible for forestry, wildlife, fisheries and agriculture/land use had internal squabbles that led to its disintegration. The powerful national economic drive, the Ministry of Agriculture and Fisheries, thus found itself disintegrating into two ministries namely Ministry of Fisheries and Ministry of Tourism (MITUR) under a new Directorate for National Conservation Areas (DNAC) with conservation areas passed on to the latter ministry for purposes of tourism in the country[27]. Although the period between mid-1990s and 2002 registered a sharp increase in protected areas from 13 % to 17 % of the country (36, 300km^2),[28] the disintegration of the Ministry of Agriculture and Fisheries resulted in a number of ambiguities and natural resources management complexities such as those explained below:

1). Conservation in Mozambique was mainly donor funded/'projectized' in terms of the three aforementioned major programs worth US$31 million. The US$31 million was meant to run natural resource management including salary supplements. As such, the division of the Ministry of

[27] See Soto, B. 2003. Protected areas management in Mozambique, Report for IUCN-SASUSG, Harare.

[28] For further discussion on CBNRM in Mozambique see Anstey, S.G. 2004. 'Governance, natural resources and complex adaptive systems: A CBNRM study of communities and resources in northern Mozambique.

Agriculture and Fisheries into two ministries implied complex division of aid projects, resources and management.

2). It became highly questionable and ambiguous in the area of natural resource and environmental management on what constituted areas and activities for tourism purposes. The decision made was that all National Parks and Hunting areas be transferred to the Ministry of Tourism.

3). The new legislation evolved during the late 1990s, the Forestry and Wildlife Law of 1999 was designed for both resources (Forestry and Wildlife) and within a single administrative structure. The 2002 Regulations to this Law are also unclear as far as differentiation of roles, responsibilities and authority in the new institutional set between DNAC and DNFFB are concerned.

4). Both the Ministries of Fisheries and Tourism had poor and limited resources, both monetary and human resources as the disintegration resulted in confusion that the main donor and aid projects felt they could not effectively sponsor processes of institutional disintegration[29].

It is important to note that though Mozambique had been using a socialist driven centralization since its independence in 1975, it finally took a new direction in the mid to late 1990s which saw the government emphasizing political decentralization and local-level resource management. This new turn was realized in the enactment of the Land Policy of 1995 and Land Law of 1997, both of which emphasized the formalization of customary rights over the most important resource, land. With these laws, local communities could now secure land and title deeds, and to an extent benefit on proceeds from the land. However, even to date the process of securing land, especially title deeds remains mired with

---

[29] Opcite.

confusion and ambiguities as the state still remains the owner of wildlife and technically of all land in the country. In fact, there are neither local-level institutions nor provision for this in legislation. This is a clear testimony that instead of devolving 'actual' rights of land and other such resources to local communities, the government remains more interested in top-down natural resource management approach which concentrates much on those programs which directly benefit or generate revenues for the state such as ecotourism, trans-boundary parks and protected areas. It therefore appears that the question of economic benefit and monopoly over authority and power remains the state's biggest preoccupation when it comes to natural resource conservation and environmental management. Such realities are not only true of Mozambique, but of other countries in the southern African region. In Zimbabwe, for example, while traditional leadership and the local communities are in control of natural resources in the rural areas, they have remained sceptical of government instruments like the Rural District Council Act of 1988 which states that all land and natural resources therein belong to the state and are managed by the latter through Rural District Councils (RDCs).

In terms of particular resources such as wildlife and forestry, Mozambique's wildlife Act of 1997 and forest policy of 1999 are the main instruments through which the government sought to create space for local community participation in natural resource management. These policy documents focus on the principles of participation of local communities, emphasizing the active participation in the management and planning processes of those who use and benefit more directly from the resources. The Mozambique's 1997 Policy and Strategy for Management, Cabinet

Resolution Number 8/97 of April 1, for example, set principles for wildlife management which include:

(a) Conserving basic resources, including biological diversity;
(b) Involving people who are dependent on forestry and wildlife resources in the planning and sustainable use of such resources; and
(c) Ensuring that [local] communities benefit from wildlife resources[30].

On the same note, Forestry Policy adopted in 1999 recommends an integrated management of natural resources that ensures effective participation of local communities, associations and the private sector, but with the latter's (private sector) involvement in natural resources management being aimed at furthering local community progress and interests. These policies were, however, in many cases hijacked by some Mozambican elites and politicians who because of greediness and corruption wanted to benefit at the expense of the rural poor. This is aptly captured by Mackenzie,[31] who notes with concern that:

The "Chinese takeaway" from the forests of Southern Africa, and in particular the Zambezia province of Mozambique is tremendous. The Chinese buyers, local business people and members of the Government of Mozambique and their forest services are colluding to strip precious tropical hardwoods from these slow growing, semi-arid and dry tropical

---

[30] See legal instrument: Law No.19/97, of October 1, Maputo, Mozambique.
[31] Mackenzie, C. 2006. Forest governance in Zambezia, Mozambique, Mozambique: Chinese Takeaway! *Final Report for FONGZA,* Maputo, pp. vi-vii.

forests at a rate that could see the resource exhausted in 5-10 years. The timber, part of it sold cheaply and illegally through corrupt means by some local business people or government officials, is exported as unprocessed logs, undermining local industry and transferring most of its potential benefits from one of the poorest countries in the world, to what is becoming one of the richest. Together with local business interests and Asian traders, these public servants constitute a "timber mafia". Instead of combating illegal logging, they are, through measures including the manipulation of forest regulations, technical information and statistics, accepting bribes and personal involvement in logging, actually facilitating and personally benefiting from this "Chinese takeaway". Mozambique government is therefore recommended to curb corruption and promote industrial development (for example, log industry) and local jobs in the long-term – and perhaps most importantly, that would fulfil a government's promises to its people.

As can be seen from the extract above, only the government (or at least some government officials) and the elite members of the society (such as some Mozambican business people and politicians) benefit from resources such as forestry which in fact should benefit majority of the rural people languishing in abject poverty. Besides Mackenzie's observation, the same has been confirmed by some critical analysts such as Lundin,[32] Hanlon,[33] and Spector *et al*[34] who observed that many of Mozambique's leaders, corrupted by the payoffs and negligence of donors and international financial institutions, are busy cashing in whatever natural

[32] See Lundin, I. 2000. *Africa Watch: Will Mozambique Remain a Success Story?* African Security Review Vol 9 No 3.
[33] Hanlon, J. 2004. *Do Donors Promote Corruption? The Case of Mozambique.* Third World Quarterly 25 (4), pp. 747-763.
[34] Spector, B, Schloss M, Green S, Hart E and T Ferrell, 2005. *Corruption Assessment: Mozambique*, USAID.

resources they can, lining their pockets at the expense of their impoverished populace. This means that Mozambique's impressive 8 % and 4.5 % (for 2004 and 2009 respectively) annual economic growth rates are fuelled by a few big private projects or individual business people and disguise continuing high rates of rural poverty[35]. This is the same point Hanlon makes when he observes that 'besides assistance from international organizations and the 4. 5% economic growth of Mozambique, the number of poor people in the country is ever increasing'[36].

Also, although the 1995, 1997 land laws, and 1997 and 1999 wildlife and forest polices formalize customary rights, this was just in principle or rather "empty words scribbled on paper" as all land and natural resources such as wildlife, fisheries and timber forest technically remained under the state's jurisdiction. The principle of state custodianship is clearly pronounced in the 2004 constitution, a revised text of the 1990 Constitution, which maintains that natural resources existing in the soil, sub-soil, internal waters, the territorial sea, the continental shelf and the exclusive economic zone, are state property, with the state determining the conditions for their use by the citizens. This can be understood to be lack of legal framework that ensures full participation of local people in the management and use of natural resources in their respective areas. Having noted this limitation for community participation, Salamao[37] argues that because of the lack of a legal framework to ensure local participation in environment

---

[35] Ministério de Planeamento e Finanças, IFPRI, Purdue University, 2004. *Poverty and Wellbeing in Mozambique: Second National Assessment.*

[36] More on Mozambique's economic growth can be found in Verdade Jornal-Newspaper, (04//12/2009)

[37] Salamao, A. 2002. *Participatory natural resources management in Mozambique: An assessment of legal and institutional arrangements for community- based natural resources management*, pp. 5.

and resource management in Mozambique, local power can be described as mere or simple privileges given and taken at the discretion of state authorities without real transfer of decision-making powers to local communities. While lack of community participation is widely criticized by many political ecology scholars, others such as Dove *et al* argue that Mozambique's reluctance to devolve "actual" rights to local communities is not surprising as in any country, it is often difficult to win state recognition of local community authority without undermining that authority in the process as there is often, although not always, a pernicious conflict between what works politically within the community and what works politically beyond it[38].

More so, while Mozambique has acknowledged the importance and need for local community members to actively participate in the conservation and management of resources in their communities, community-based natural resource management (CBNRM) especially organic in the country has not been well established or rather legitimized at national level. By legitimacy, I mean 'any behaviour or set of circumstances that society defines as just, correct, or appropriate'[39]. In the context of this study, society is the local population affected by the conservation strategy imposed on them or negotiated with them. This issue of CBNRM will be pursued in greater detail in chapter 5. In this chapter, it suffice to say there are no bold initiatives taken towards CBNRM even as externally initiated as in other southern

---

[38] For further discussion on devolution of rights read Dove, R. M., Sajise, E. P., and Doolittle, A. A. 2011. *Beyond the sacred forest – Complicating conservation in south-eastern Asia,* pp. 28.

[39] For more on the conceptualization of legitimacy read Brechin, S.R., P. R. Wilshusen, C. L. Fortwangler and P.C. West. 2003. *Contested Nature,* pp. 14.

African countries such as Namibia, Zambia, Zimbabwe and Malawi, among others. In Zimbabwe and Malawi, for example, CBNRM has been legitimized and nationalized in the names of Community-Based Fisheries Management (CBFM) and Communal Areas Management Program for Indigenous Resources (CAMPFIRE) respectively. In Zambia and Namibia, CBNRM has been legitimized and nationalized in the names of Administrative Management Design for Game Management Areas (AMANDE) and Namibia Wildlife Trust with community game guards and later Namibia Wildlife Management, Utilization and Tourism in Communal Areas. Yet in Mozambique, CBNRM remains more of a theory that has not up till now been fully implemented and nationalized the country-over, although there are some instances in the central and northern provinces of the country where the communal members have organized themselves or have been organized by civic organizations and lobbied to protect their own forest and other resources therein. As such, there are only instances of projects in some provinces in the country that can be characterized as CBNRM if we are to consider who initiates the projects, the objectives of the projects, and methodologies used in executing activities of the projects. The examples of such projects include Ancuabe (in Cabo – Delgado province), Chipanje Chetu (in Niassa province), and Sanhote (in Nampula province), and Pindanyanga (in Manica province).

The background and historical overview of natural resource management paraded in this chapter partly explains why even though natural resource conservation and environmental management in Mozambique took a new twist after independence, has failed to disentangle and do away with the crisis the country inherited from the Portuguese colonial government, but rather extended or even exacerbated the problems.

In concluding this chapter, I underscore that the environmental and natural resources conservation in Mozambique from the colonial era (Portuguese regime) through the present have been critically and historically examined. The examination was made to unravel and bring to light the trend, and relations that existed in environmental and natural resource management through time for readers to understand what we mean when we say "Mozambique's environmental and resource conservation is in crisis", and to draw lessons from the previous conservation efforts and attempts. On that note, it has been emphasized that success of natural resource conservation does not only depend on the availability of resources, but on how power and authority are distributed within the country and conservation systems. Power wrangles, poor state legislation/legal instruments for promotion of sustainable conservation, and internal conflicts within the bodies responsible for conservation of natural resources have been highlighted as the major causes and fosters of conservation crisis in Mozambique.

It has also been noted from the history of natural resources conservation and environmental management paraded above that since colonial period (in Mozambique) to the present, community participation in issues of resource conservation has been minimal. This partly explains the reason why in Mozambique to date, there is no single accepted definition of what is meant by *'Maneio Comunitario dos Recursos Naturais'* (Community-Based Natural Resource Management) (CBNRM) as in other countries in the region such as Malawi, Namibia and Zimbabwe. In view of this observation, it has been argued that although there are initiatives in some rural areas of Mozambique which can be considered as CBNRM,[40] the CBNRM is not a fully legitimized and nationalized conservation strategy as in the aforementioned countries. Also, the so-called CBNRM in Mozambique is characterized with diverse principles depending on who is leading the project, unlike in other countries in the region where the principles that guide such programs are more or less the same throughout the country.

In the light of the sad history of natural resource conservation in Mozambique paraded in the preceding paragraphs, the chapter concludes by noting that there is a long standing conservation crisis in the country that requires total commitment by the government and all parties concerned in order to unlock and resolve it.

---

[40] See Nhantumbo, I., Chonguisa, E., and Anstey, S. 2003. Community-based natural resource management in Mozambique: The challenges of sustainability, Report to IUCN-SASUSG.

Chapter 2

Environmental Malpractices in Mozambique: A Case Study of Xai-Xai Communal Area, Gaza Province

Recent climate of the world is now known to have fluctuated frequently and extensively in the three or so million years during which humans have inhabited the earth[41]. Recent studies have confirmed that the world is undergoing serious climate change as global temperature is ever increasing. While the bulk of the changes that the global world is undergoing have nothing to do with human intervention,[42] the Third Assessment Report of the Intergovernmental Panel on Climate Change,[43] reveals that the global average temperature will increase by 1.4° C to 5.8° C between 1990 and 2100 if the levels of emissions are not reduced. According to the same report, though the bulk of global changes are believed to be natural, the emission of carbon dioxide and, therefore, increase in temperature in the atmosphere is largely attributed to the anthropogenic activities especially the global use of fossil fuels. On the other hand, developing countries especially in Africa are even more vulnerable due to their low adaptive capacities and their dependence on rain-fed

---

[41]See Goudie, A. 1983. *Environmental change,* Clarendon Press: Oxford.

[42] See Goudie, A. 1990, *The human impact on the natural environment,* 3rd ed, Basil Blackwell Publishers: UK, pp. 262.

[43] More on climate change can be found in Intergovernmental Panel on Climate Change (IPCC), 2001. Third Assessment Report: Climate Change TAR, Available at: http://www.grida.no/publications/ipcc_tar/.

agriculture,[44] and small-scale subsistence agriculture for their livelihood security[45]. The impacts of climate change in Africa as elsewhere are generally manifested in human health and in the agricultural sector worsening the existing levels of poverty and undermining all development efforts in the continent. In Mozambique, for example, since about 80% of the population depend directly on land and other natural resources, the effects of climate change and variability are likely to have tremendous consequences on the rural communities and the national economy in general. The main sectors likely to be impacted by climate change according to Mozambique's initial Communication to UNFCCC[46] include: Agriculture and food security, water resources, costal resources, biodiversity, human health, loss of life, erosion, land degradation, sea level rise, natural disasters, salt intrusion, crops, ecosystems, property, human and animal habitats, outbreaks of pests and diseases, displacement of people, and destruction of infrastructure (communication network, schools, hospitals and houses).

In view of these forecasted problems, environmental protagonists have relentlessly championed "good" and

---

[44] See FAO report. 1995, *World Agriculture: Towards 2010. An FAO Study,* Ed. N. Alexandratos. FAO, Rome, Italy, pp. 481.

[45] Rockström, J. 2000. Water resources Management in smallholder Farms in Eastern and Southern Africa: An Overview. Phys. Chem. Earth (B), Vol 25, No. 3, pp. 275-283.

[46] For further discussion on climate change see Mozambique First National Communication to the United Nations Framework Convention on Climate Change (UNFCCC), 2006, Submitted to the Secretariat of the Convention in 2006. Maputo – Mozambique; See also Bambaige, A. 2008. National Adaptation Strategies to Climate Change Impacts: A Case Study of Mozambique, World Human Development Report Office.

sustainable environmental practices as indispensable catalysts for fostering human health and fighting global warming in general. However, these protagonists have championed environmental practices but without critically reflecting on and taking stock of the deep seated structural constraints (poverty, high levels of illiteracy, lack of environmental and agricultural expertise among the rural populace), ideological and political motivations that accompany good environmental practices adoption and use in resource constrained environments[47]. Not surprisingly, several studies on environmental management evolve and are guided by the assumptions that increasing access to information particularly secondary sources/material like books and Information and Communication Technology (ICT) will inevitably scale up access to information and accelerate the production of knowledge[48]. This understanding has been accentuated by the role of the Internet and the World Wide Web in "democratizing" the production, dissemination and access to information. The interventions at national and educational institution levels have been guided by the unsubstantiated assumption that increasing the ICT infrastructure (internet and communication networks, computers, electricity supply) and other secondary materials like books will automatically improve human knowledge on environmental management without paying sufficient attention to other structural constraints that hinder access to the information like the ICT skills gaps of academics and laypeople, crippling poverty

---

[47] See Federal Republic of Nigeria National Policy on Environment (FRNNPE) 2004. Nigeria's National Policy on Environment, Nigeria.

[48] See Heurtin and Licoppe. 2001. Handbook of Research on User Interface Design and Evaluation for Mobile Technology, Taylor and Harper.

levels, high levels of illiteracy, socio-cultural barriers like negative perceptions towards new technologies such as pedagogical use of ICT and student limited communicative competencies. In the light of this observation, it can be argued that the empowerment of peasant farmers (farmers at grassroots) and other rural dwellers should be in the forefront on environment and natural resources conservation issues that affect their communities. Put it differently, rural people should be made *prosumers*[49], that is, environmental managers and users of resources in their own communities. Prosuming conservation, thus, is critical as it is likely to promote sustainable conservation of the environment and other natural resources.

While the aforementioned points are critical to Africa and the world in general, in Mozambique, the recommendation to adopt presuming conservation is more urgent than ever. Besides, the causes and impacts of environmental changes are also greatly visible, hence one reason (among many others) why this study has adopted Mozambique, and in particular Xai-Xai communal area as a case study. Mozambique is among the most disaster prone countries in the world – combined with high levels of abject poverty, high levels of illiteracy and weak national institutions. The occurrence of natural disasters such as floods, cyclones, drought and earthquakes has consistently impacted significantly the people's lives and the national economy: they have created a life-threatening situation for the Mozambican people.

---

[49] The idea of presumption was borrowed from professor Busher, B. who presented a paper "Prosuming conservation" to the Department of Sociology, University of Cape Town, (25/02/13).

According to United Nations report[50], humans are the most endangered species in Mozambique. Other threatening environmental problems include the loss of the nation's forests through deforestation and illegal logging, and human induced veld fires. Making a case for the latter, Mozambique lost 7.7% of its forest and woodland between 1983 and 1993 alone[51]. In 2008, 'veld fires in the central provinces of Mozambique allegedly started by local farmers extending their fields destroyed 16,000 hectares of arable land'[52]. Besides the loss of forests by Mozambique, the purity of the nation's water supply is also a significant issue. Surface and coastal waters have been affected by pollution. Mozambique has 100 cu km of renewable water resources. About 89% is used in farming and 2% for industrial purposes. Only 81% of the nation's city dwellers and 41% of the rural population have access to pure drinking water. As of 2001, 13 of the nation's mammal species and 14 bird species were endangered and about 57 plant species were threatened with extinction. Endangered species in Mozambique include the green sea, hawksbill, olive ridley, and leatherback turtles[53].

The abovementioned factors make Mozambique susceptible to negative environmental consequences and human-induced climate change that could even worsen the

---

[50] See United Nations Environment Program (UNEP)report, 1999. GEO-2000 global environmental outlook, Nairobi.

[51] See United Nations Environment Program (UNEP) report, 2002. Vital climate graphics Africa (available at www.grida.no).

[52] For more on green revolution in Mozambique see Africa News Network, 2008. 'Mozambique green revolution will depend on small scale farmers':http://www.afrol.com/Mozambique.BBCNEWS/http://news.bbc.co.uk/go/pr/fr/-/2/hi/africa/7601114.stm.

[53] See United Nations (UN) report. 2002. UNDAF II 2002-2006 (available at: www.unsystemmoz.org/).

gravity of the already existing problems. Projected consequences for Mozambique consist of sea level rise (SLR), water scarcity, floods, agriculture and food insufficiency, increased soil erosion, increased wind speed, loss of biodiversity and habitats and new pressures on human health and national economy. It is estimated that as much as 25 per cent of the population faces a high mortality risk from such events, and Mozambique ranks as the second most geographically exposed country in Africa[54]. Climate change effects have the potential to increase this risk in the future, and can easily undermine development efforts and increase vulnerability of poor people who depend disproportionately on the environment for their livelihood – which may lead to social, economic and, political unrest in the country. These projected effects, thus, underscore the necessity for serious consideration of sustainable resource use and environmental conservation in poverty alleviation interventions, irrespective of climate change concerns. These are all issues that demand careful and immediate attention (from both the government and nongovernmental organizations), if an environment ready for sustainable development is to be guaranteed in Mozambique and of course beyond its borders; hence the present study is timely and relevant.

Background to the study problem

While the discourse on environmental management has, to a larger extent, gathered momentum in Southern Africa, Mozambique included, in the last couple of years, most of the issues raised have centred on solving the problem at national

---

[54] See the 2009 United Nations Development Program (UNDP) Human Development Report on Mozambique, Available online at: http://hdrstats.undp.org/en/countries/data_sheets/cty_ds_MOZ.html

and global levels without seriously considering the contribution(s) of the local or 'grassroots communities'. Issues of global warming, depletion of the ozone layer, acid rains, famine and other such natural and anthropogenic induced disasters have dominated most academic discourses and researches and very little has been written about the rural dwellers' place and relevance in natural resource conservation and environmental management. In Mozambique, and in particular Xai-Xai communal area, what is even more worrying is the fact that it is still the outsiders/externals such as Non-governmental Organizations (NGOs) who are driving the environmental management initiatives (though only does this occasionally) and the community dwellers/local farmers are the passive recipients. In fact the local communities are in most cases not included in the decision-making processes about what has to be done to arrest environmental degradation and to promote sustainable conservation in their areas. It is this kind of background which prompted me to come up with the following research question for this study: 'Could discourses on environmental management at national level alone impact a positive change of behaviour on the rural population and improve the future of Mozambique's environmental management strategies?'

Despite the lack of specific studies carried out in the Xai-Xai communal area on prevailing environmental problems, it has been asserted that the prevalence of natural disasters and other such environmental problems tend to be worsened by environmental malpractices by people. Recent study on Mozambique's climate in general shows that if no action is taken as a matter of urgency the effects of climate change in the near future will be hard to bear as they include the following:

31

♦ Increase of the mean air temperature by between 1.8 and 3.2 °C;

♦ Reduction of annual rainfall by 2 to 9%;

♦ Increase of the solar radiation from 2 to 3%; and

♦ Increase of the evapo-transpiration by between 9 to 13%[55].

The above is a clear testimony that urgent measures should be put in place to curb environmental problems and lessens the gravity of natural disasters such as floods which in fact have become a perennial phenomenon in both the rural and urban areas of Xai-Xai district. I was therefore motivated to carry out research on natural resource conservation and environmental management in Xai-Xai communal area so as to shed more light on this rather neglected area. Environmental malpractice is considered to be universal in Mozambique and hence this research describes how such malpractices manifest and suggest how the malpractices could be dealt with now and in the future.

Research objectives

Could results from a critical examination of environmental malpractices be used to impact a positive change of behaviour on the rural population and improve the future of Mozambique's natural resources and environmental management strategies? Most researchers on natural

---

[55] See Mozambique First National Communication to the United Nations Framework Convention on Climate Change (UNFCCC), 2006, Submitted to the Secretariat of the Convention in 2006; See also Bambaige, A. 2008. National Adaptation Strategies to Climate Change Impacts: A Case Study of Mozambique. World Human Development Report Office.

resources and environmental issues on Mozambique[56] are guilty of trying to solve resources and environmental problems only at national level, without directly engaging the local communities in such discourses. They seem to be unaware or rather forget that the rural communities participate significantly in the creation of resources and environmental problems they face. As such, the aforementioned communities, if engaged, can play a significant role in combating the problems because they know (partly or in full) how the problems are created and impact on them as communities. In view of this observation, it is argued in this book that the history and discourses of natural resource conservation and environmental management in Mozambique make a sorry reading especially with their failure to document, deliberately or otherwise, the place, role and contribution of rural communities in resources and environmental conservation and management systems. That said, this study serves as a corrective to the reality in Xai-Xai and beyond that the rural dwellers are made 'back seaters' [by the government and other such organizations] in the whole attempt to redress the environmental problems of their own communities – a move that if uncorrected would result in the mounting of 'environmental crisis in the country' and ultimately contribute to the global environmental crisis.

Geographical description of the study area

---

[56] For further discussion on agriculture and conservation in Mozambique see Wisner, B. 1977. Agriculture *in Mozambique – Science for People (London)* 34 [pages unknown]; See also FAO. 1997. Irrigation potential in Africa: A basin approach. *Land and Water Bulletin No. 4.* Rome; See also World Bank, 2009. *Development and Climate Change,* The World Bank group at work. Washington DC 20433.

The study elaborated in this chapter was conducted in Gaza province of Mozambique, Xai-Xai district using Xai-Xai communal area as a case study. Xai-Xai is located close to the Indian Ocean, on the Limpopo River, which run through the district emptying into the Indian Ocean near Xai-Xai city. It is 200 kilometres (120 mi) from the capital, Maputo, and is in a wide, fertile plain where rice is grown. Xai-Xai has a population estimate of 116,343[57]. The beach of Praia do Xai-Xai (Xai-Xai beach) is the major tourist attraction in the district. The rural area stretches from the southern part of the city to the north and to the south eastern part of Gaza province along the coast of the Indian Ocean. The relief of the area is characterized with steep slopes and isolated sand dunes on the eastern side. The altitude of the area is about 500 – 1400m above sea level but with some low lying areas below sea level. The area has warm to hot summers and cool winters. The mean annual temperature is about 30°C. The rainfall though falls throughout the year, depending on the year and influence from the sea, is mainly conventional and occurs between November and March. The climate of the area is the tropical type as it includes one rainy season and one dry season; it is hotter during dry season and cooler during wet season. The mean annual rainfall for the area is about 500 – 750 mm per year with vegetation consisting of a Tropical dry savannah. Below is the map of Mozambique

---

[57] See World Gazetteer, *2007,* ^ *a b c* "Mozambique: largest cities and towns and statistics of their Population". Available at:http://worldgazetteer.com/wg.php?x=&men=gcis&lng=en&des=wg&srt=npn&col=abcdefghinoq&msz=1500&geo=-153. Retrieved 2008-06-17.

showing the geographical location of the case study (Xai-Xai communal area).

Figure 1: Xai-Xai Communal area, Gaza, Mozambique

The next section takes a look at the methods used to carry out this study.

Methodological issues

As part of my research design, I relied on literature studies, content analysis, field observation and in-depth interviews. The study was carried out between April and July 2010 using a selected sampled of 50 people (30 female and 20 male). The sample size of 50 was considered sufficient in providing the general resources exploitation and environmental perceptions of the people of Xai-Xai communal area. The data collection procedures which included content analysis, observations, in-depth and structured interviews were considered suitable for a study such as this because they allowed a deeper and comprehensive understanding of what was going on in the area. I observed the physical environment and resources therein as well as the environmental malpractices in the chosen area. The field observation was used to ascertain the project location and what really happened at local level. To supplement the field observation information, interviews were conducted with the intention to get more information on the possible causes of environmental problems and how they can be dealt with at local/community level. The people who participated in the study were from different societal classes, ranging from the educated to the uneducated, and from the working to the non-working classes. The respondents were drawn from different societal classes with the hope for obtaining a balanced research result that could be representative of the whole studied area. They ranged from 13 to 60 years. This age group was considered appropriate for the study given that most of the people involved in direct use of natural resources and in environmental (mis)management in Mozambique are between the aforesaid age ranges. More women than men were sampled for the mere reason that it is generally believed women are in most cases less represented and sometimes misrepresented in social science researches.

I administered questionnaires with both open and closed items (open questionnaire and closed questionnaire) to the participants in the different areas they were found. The open questionnaire was used as it enables the respondent to reply as she [he] likes and does not confine the latter to a single alternative[58]. In fact it evokes a fuller and richer response as it goes beyond statistical data into hidden motivations that lie behind attitudes, interests, preferences and decisions. Open questionnaire possibly probes deeper than the closed questionnaire. On the other hand, the closed questionnaire was used because it facilitates answering and makes it easier for the researcher to code and classify responses especially in this case where a large number of questionnaires were to be dealt with. Both questionnaires were used because in practice, a good questionnaire should contain both open and closed forms of questions so that responses from the two forms can be checked and compared[59]. The participants responded to questionnaire items individually and participation was voluntary. Participants were assured of the confidentiality of their responses and were asked not to identify themselves by names. Collected data were tabulated to show frequencies before being subjected to evaluative analysis. The Tables 1

---

[58] Behr, A. L. 1988. *Empirical Research Methods for the Human Sciences* (Second Edition), Durban Butterworths.
[59] Ibid.

and 2 respectively contain details of the people participated in the study and the data that was gathered during the study:

Table 1: Participant demographics

Occupation	Gender	
	Male	Female
Cattle headsman	2	2
Village head	3	2
Fisherman	2	2
Traditional healer	1	2
Peasant Farmers	4	8
Charcoal producers	3	2
Students in public education	2	5
Agritex Officers	1	3
Firewood sellers	2	4

Table 2: Responses to closed questionnaire items

ITEM	RESPONSES		
	Agree	Disagree	Uncertain
1.There are certain species of plants that should not be cut down or burn	7	43	0
2. Each community should have a cultural village or a protected area	6	43	1
3. Fishing should be controlled i.e. done with licensed people and not done any time of the year	40	6	4
4. Veld fires should be discouraged and avoided at all costs	36	10	4
5. Crop rotation is better than monoculture	40	6	4
6. Charcoal production promotes deforestation and land degradation	30	15	5
7. Fire wood selling promotes deforestation	15	30	5
8. Crop residues should not be burnt but ploughed back after harvesting	26	20	4
9. Animal hunting should be controlled i.e. done by licensed people and not done any time of the year	40	7	3
10. Each community should have paddocks/reserved pastures for livestock	32	13	5
11. All community members should take part in the prevention of soil erosion and conservation of natural resources	38	10	2
12. Villagers should look for alternative forms of energy like gas, electricity and solar instead of	34	15	1

39

relying solely on firewood and locally produced charcoal.			
13. Villagers should find alternative building material like cement bricks instead of solely depending on the locally available ones	35	14	1
14. Each community should have Agritex officers to help farmers	46	4	0
15. There should be government and non-governmental initiatives on environmental management issues	45	4	1

Results and discussions

The study showed both positive and negative perceptions of the local people on natural resource use and on what environmental (mis)management means. For instance, there were mixed feelings with regard to whether villagers should look for alternative sources of energy to substitute firewood and locally produced charcoal. Similar results were obtained on whether villagers should look for alternative building material instead of relying solely on the locally available ones such as thatching grass and poles. Reasons given varied across the studied area but the major one was that alternative sources of energy and building material are more expensive than the locally available ones. This suggests the government should devise ways to ensure that prices of alternative sources of energy such as gas and electricity, and building material (such as treated poles and asbestos) are at par with those of the locally available ones. On whether fishing and hunting should be controlled, an overwhelming majority (40 out of 50) of the respondents agreed, thus showing support for the protection of some fish and animal species, especially the endangered species. And, on whether veld fires should be used as means of clearing and preparing farm lands, majority (36 out of 50) agreed that veld fires should be avoided at all costs. This finding auger well with results from a recent study on Mozambique's climatic conditions by Mozambican Initial Communication to the UNFCCC[60] which notes that if no action is urgently taken to control climatic changes there will be increase in mean air temperature, reduction of annual rainfall and increase in solar radiation. It also relates to my recent studies (see Chapter 3, this volume) on green revolution program in Mozambique, where similar findings were reported. In these studies, it was revealed that between 2008 and 2009 in Mozambique, particularly in the central provinces of Manica, Sofala and Nampula, veld fires claimed more than 20 lives, violated non-human animal rights and destroyed property as well as more than 16,000 hectares of arable land. In the light of these touching findings, it becomes imperative for the Mozambican government to reconsider the use of veld fires in farm areas, to incorporate environmental ethics that respect animal and other such entities' rights in its green revolution program. It should be remarked, however, that the present study showed that there are a minority (10 out of 50) in favour of the status quo or 'business as usual' as far as the use of fire is concerned. Their beliefs and attitudes concerning man's place in the natural world (often seen as nature's place in man's world) and the use of fire as part of

---

[60] See Mozambique First National Communication to the United Nations Framework Convention on Climate Change (UNFCCC), 2006; See also Bambaige, A. 2008. National Adaptation Strategies to Climate Change Impacts: A Case Study of Mozambique, *World Human Development Report Office.*

land preparation (for farming) are so deeply woven into the fabric of traditional culture that they have accepted them and even unconsciously. This is contrary to the reality that our poor, battered, plundered and polluted planet can no longer endure a continuation of 'business as usual'. In this light, it is critical for the Mozambican government to encourage (through civic education) the rural people to do away with the 'unruly' belief that the ecosystem is not understood or even recognized as a system and that the earth and its wilderness is too vast to be damaged by voluntary human choice. In our own time we need to re-validate the symbiotic relationships and interdependence between us (humans) and the [natural] environment, for in our own time, with knowledge has come power, and with both knowledge and power, we have lost our senses to the extent of damaging the environment on which our lives depend.

Concerning the government and other nongovernmental organizations' role on resources and environmental management, majority (45 out of 50 respondents) indicated that there was need for the government to empower people with knowledge and sustainable ways of managing their environment and natural resources alongside their farming activities. This finding concurs with results from a recent study by Africa News Network[61] which revealed that poor farming (in Majune, Mozambique) could wipe away the scenic beauty of the area even before it can realize its potential as a successful ecotourism product, as evidence of indiscriminate

---

[61] For more on green revolution campaign in Mozambique read: Africa News Network, 2008. "Mozambique green revolution will depend on small scale farmers", Mozambique Afrol News, Also available at: *http://www.afrol.com/Mozambique*.BBCNEWS/*http://news.bbc.co.uk/go/pr/fr/-/2/hi/africa/7601114.stm.*

burning of the bush are common and nobody (including the government) seems to care. Thus due to lack of government will/initiative, many people by default or otherwise are reluctant to actively take part in good resources and environmental management, hence making the state an "accomplice" in all problems that have to do with natural resource conservation and environmental mismanagement. Though there are visible efforts by the government to electrify the Xai-Xai rural area, the pace at which the project is being implemented is too slow and the energy too expensive for the ordinary people to afford. As such, 99% of the population continue using charcoal and firewood for cooking and heating. For them, charcoal and firewood are relatively cheaper as compared to alternative sources – electrical energy and gas. However, the continued use of charcoal and firewood will always impact negatively on resource use and the natural environment in general as it encourages deforestation, erosion, pollution (during charcoal production) and general environmental degradation. More worrisome is the realization that no efforts (in the studied area) have been made so far to encourage people to use other forms of energy that are pollution free and do not harm the natural environment. This is likely to change rainfall pattern in the area as has been revealed by the increased annual occurrences of the natural disaster, floods, in the last few years. The 2010 – 2011 and 2013 floods, for example, hit hard on the Xai-Xai area resulting in human life, property and other such valuable resources being lost. It is also likely that hotter and drier conditions would widen the area prone to desertification (the arid zone of nearby districts like Massangena, Chicualacuala, Chigubo, Mabalane and Massingir) into the Xai-Xai communal area. It goes without mention that the socio-economic costs of desertification would be tremendous and overwhelming, especially considering that Mozambique is one of the poorest countries in the world. Having discussed study results, environmental malpractices observed in Xai-Xai in terms of their possible causes and impact are examined.

Possible causes and impacts of environmental malpractices in Xai-xai communal area: A closer look

♦ *Poor fishing habits*

As described above, Xai-Xai is along the coast of Indian Ocean. Its shores are among the most spectacular in the world with a substantial number of the currently endangered species – whale sharks. However, the shores are suffering from illegal and over- fishing (unregulated fishing like fishing out of season) by both local and foreign fishermen mainly from China). As Leo Buscaglia,[62] a tourist observed and indeed lamented:

Mozambique has fallen into a common trap of many poor African nations, and has allowed China to have full rights to their waters. This allows the Chinese to fish by their own laws in Mozambique and the local government will turn a blind eye to any of their malpractices, as long as the Chinese build and fix some important pot-hole strewn roads in the country. We saw various Chinese forklifts and trucks tarring the road stretching north-eastwards from Xai-Xai city whilst we were driving through the area.

While it is a fact that there are only 1000 Whale Sharks in the world and about 300 of them are in Mozambique, these

---

[62] For a full report on Chinese operations in Mozambique read: Buscaglia, L. 2010. "Milk and Honey in Mozambique". Also available at: http://www.milkandhoney2010.com/201004/mozambique.html

allegedly fall prey from both the local and foreign poachers/unregulated fishermen[63]. This in no doubt is depriving the country's economy and of course future generations who are unlikely going to be privileged to have a glimpse to these rare and endangered creatures in their natural environment. This is because a substantial number of the currently endangered species (whale sharks) might be lost as coastal habitats are lost and invaded by poachers or unregulated fishermen. On this note, Buscaglia's[64] words should be seriously considered when he noted: 'we remain with little hope, that this doesn't change because of China's presence in the region and that many people in the future still get to enjoy these beautiful beasts'.

◆ *Poor farming methods*

As reported by FAO,[65] in the productive coastal zone in Gaza province (Bilene-Macia, Xai-Xai, Manjacaze, Chokwé, Guijá and Chibuto), conditions are "relatively" favourable for agricultural production (of crops such as maize, manioc, rice, *mafura*/castor beans, tobacco, cotton, sugar cane, banana and fruits like mangoes, oranges and cashew). However, poor methods of farming by local subsistence farmers are not only exacerbating food insecurity in the area, but are also posing serious problems to environmental degradation as many farmers practice monoculture, and cultivate along streams and down slopes. Such poor farming methods put the entire area to high risks of soil erosion and siltation of the nearby Limpopo River. This calls for soil conservation measures to

---

[63] Ibid.

[64] Ibid.

[65] For further discussion on agricultural production in Mozambique read FAO. 1997. Irrigation potential in Africa: A basin approach. *Land and Water Bulletin No. 4.* Rome.

manage and conserve water resources, 'as soil erosion is causing high silt loads and turbidity in the Limpopo River and its tributaries, affecting water treatment and the storage capacity of[66] the River. Xai-Xai has some tributaries that feed into the Limpopo River like the Rio Changane (Changane River) and Rio dos Elefantes (Olifants River), yet as highlighted above subsistence farmers in the area practice stream bank cultivation, steep slope cultivation and seem to be doing nothing to control soil erosion. One of the interviewees, a teacher from Maxelene village (in Xai-Xai rural area) affirmed this when she commented: "In this area many people do nothing to prevent soil erosion because they are not aware of the importance of doing so as most of them are either uneducated or receive no incentive from the government to do so".

♦ Veld fires

Veld fires are 'blazes that get out of control and devastate extensive tracts of forest, grassland, wildlife and other natural resources as well as injure and kill people and destroy their properties'[67] (also see chapter 3, this volume). While some fires are caused by lightning which on average strikes the land surface of the globe one hundred thousand times a day,[68] others by unknown causes as well as sparks produced by

---

[66] See Medicins Sans Frontiere (MSF)/(Doctors Without borders), 1998. 'Attacks as told by victims MSF'. *MSF Article,* Retrieved May 10, 2010.

[67] For a working definition of veld fires read Mkwanazi, H. 2007. Veld fire Campaign in Lupane. *Environment Africa*, Lupane.

[68] For more on lightning striking on earth read Yi-fu Tuan. 1971. Man and nature, *Commission on College Geography Resource Paper, 10.*

falling boulders and landslides,[69] and some result from spontaneous combustion in those ecosystems where heavy vegetal accumulations may become compacted, rot and ferment thus generating heat,[70] human beings are responsible for 95% of forest and veld fires in southern Africa,[71] as natural fires (not influenced directly by human beings) have become rare. Xai-Xai like other rural areas in southern Africa, and in particular Mozambique is suffering constant anthropogenic veld fires. In the recent times, the fires have become more and more common in most parts of rural Mozambique due to government's campaign for the country to lead green revolution in sub-Saharan Africa. The call for the country to lead the green revolution program has resulted in hasty and unplanned form of agriculture by poor peasant farmers in the countryside (also see chapter 3, this volume). In Xai-Xai, peasant farmers who lack mechanized farm implements have resorted to veld fires as one way of clearing the land. This has had serious negative impacts on the physical environment-both fauna and flora besides that veld fires contribute to a significant proportion of land degradation (as I observed in Xai-Xai) and the release of greenhouse gases to the atmosphere. Furthermore, fires destroy property and resources needed for immediate use over the dry season, crops, firewood, water sources and grazing land. Frequent veld fires may also have long-term

---

[69] Booyson, P. and Tainton, N. M. (eds). 1984. *Ecological effects of fire in South African ecosystems*, Springer-Verlag: Berlin.

[70] Vogl, R. J. 1974. Effects of fires on grasslands, In Kozlowski, T. T. and Ahlgren, C. C. (eds), Fire and ecosystems, Academic Press: New York, pp. 139-94.

[71] For more on veld fires in southern Africa see Mkwanazi, H. 2007. Veld fire Campaign in Lupane, *Environment Africa*, Lupane.

effect on the reproductive capacity of important veld products such as vegetation and humus.

More so, the fires can destroy biological diversity which in itself has many benefits for human beings as its different kinds and species contribute in providing agricultural, fishing and livestock services, opportunity for scientific research and act as symbol of cultural heritage. Some flora and fauna species with their genetic components help in developing medical, agricultural and industrial sectors. Additionally, they provide essential daily needs for the life of many local communities, biological diversity support, and ecotourism with its great economic potential. Uncontrolled anthropogenic veld fires should be, therefore, avoided at all cost as their [negative] impact on the environment, natural resources, and human life is always difficult to bear. Below is photography showing anthropogenic veld fires in rural Xai-Xai.

Picture showing anthropogenic veld fires: Courtesy – Munyaradzi Mawere

♦ Deforestation

Besides the use of veld fires to expand farmlands, people in Xai-Xai communal area cut down trees for firewood and charcoal production. Approximately 99 % of the people in the communal area use charcoal and firewood for cooking and heating on almost daily basis. Only about 1% uses other sources of energy such as gas and electricity for cooking and heating. The charcoal is produced locally though due to scarcity of trees in the Xai-Xai communal area the bulk of it now comes from other districts in the province such as Massingir and Guija where trees are still an abundant resource. Charcoal production and use of firewood has contributed quite substantially to the deterioration of the natural vegetation cover, and to soil erosion in Xai-Xai communal area. A foreseeable negative consequence of these current impacts of deforestation is a further vulnerability of Xai-Xai area to natural disasters such as cyclones, floods, droughts as well as a change in rainfall pattern and increase in atmospheric temperature.

Confronting the jeopardy: The way forward

The conservation of natural environments and biodiversity has become a fundamental global concern and is recognized increasingly as an economic, socio-cultural issue. The findings of this research show that Xai-Xai's resources and environment in general are in danger due to a number of threatening environmental malpractices such as uncontrolled fishing, veld fires and deforestation, among others. There is, therefore, need for immediate action if further damage is to be avoided and sustainability achieved.

In view of the Xai-Xai environmental malpractices discussed in the preceding section, the study advocates empowerment of the rural community and active participation of the latter in managing their own natural environment. It is essential for policy-makers to realize that environmental problems can never be solved at the national and even global level if they are not tackled from the grassroots, that is, at community level. People centred or farmer-driven approaches should be adopted. In this respect, it is important that the needs and aspirations of potential beneficiaries/communities be taken into account during the initial planning of any environmental intervention strategy. Inherent capabilities, intelligence, knowledge and responsibility of rural people will have to be taken into consideration in both planning and implementation. This is because capabilities, intelligences, knowledges and responsibilities of the rural people can best be realized and meaningfully exploited if the people concerned are first empowered before being made to participate. The principle of empowerment states that all people should participate in every activity (problem) that directly affect them because it is the democratic right of all people to do so,[72] and participation means having power[73] to take part in a particular activity. According to this concept, participation is the natural result

---

[72] Wignaraja, P.A., Hussain, A., Sethi, H., and Wignaraja, G., (eds). 1991. *Participatory Development,* Oxford University Press.

[73] For more on sustainable development read Tascconi, L. Tisdell, C. 1993. Holistic Sustainable Development: Implications for Planning Processes, Foreign Aid and Support for Research, *Third World Planning Review,* 14: 4.

of empowerment. The concept of community participation is viewed as a basis for project success. The World Bank thus defines participation as 'a process through which stakeholders' influence and share control over development initiatives, and the decisions and resources which affect them.'[74] The objectives of community participation are the following: building beneficiary capacity; increasing project effectiveness; improving project efficiency; project cost sharing and empowerment. In this sense, empowerment is not a means to an end but is the objective of development right from the community level. Empowerment entails more than having the power to make decisions. It demands the knowledge and understanding to make correct decisions by those that are directly affected. This is because communities cannot make wise decisions if they do not have the required information/knowledge. The support organizations and government agencies are required to be sources of information and should be a channel of information to the communities so that they will be able to make right and informed decisions when they interact and exploit resources from their respective environments.

In terms of poor farming methods, subsistent farmers can be empowered through civic education. This could be done through local government. The local government can achieve this by making use of Agriculture Extension Officers. The findings of this research have shown that there are no workshops/meetings held so far between Agriculture Extension Officers or the local government and the rural populace on environmental issues. One of the respondents, a village headman in Xai-Xai communal area, for example,

---

[74] See *World Development Report: Making Services Work for Poor People*, 2004, Washington DC: World Bank.

revealed that since he became a headman in 1994, his community has never had any meeting on environmental management with government authorities or qualified Agriculture Extension Officers. This confirms reluctance on the part of the Mozambican government and non-governmental organizations to deal with environmental issues at grassroots levels though a lot is being done at national level and even more at global level. In view of these findings I contend that the government should make sure that qualified Agricultural Extension Officers constantly hold environmental meetings/workshops with the rural people advising them of good environmental conservation practices (such as prevention of soil erosion and afforestation with emphasis on planting indigenous species, in particular nitrogen-fixing *Acacia* spp, and others such as *Albizia* spp, *Combretum* spp, *Zizyphus* spp, *Terminalia sericea*, and *Sclerocarya birrea*). Mozambique has the potential to develop agro-forestry, and offers a wide range of applications[75]. This potential, however, can only be fully developed and realized if the people at grassroots are also involved in contributing to the national development of such projects.

Good farming methods such as conservation farming could also be encouraged especially as a strategy to control erosion in Xai-Xai rural area. Conservation farming involves minimum disturbance of the soil surface by using an ox-drawn ripper tine to open the planting furrow. It has been

---

[75] See Lulandala, L. L. 1991. Agro-forestry potentials for Mozambique. UNDP/FAO/GOM Project MOZ/88/029. Maputo; See also Southern African Development Community (SADC), 1996. Short term consultancy in agro-forestry diagnosis and design for some SADC countries. Vol. 2 Botswana, 4 Mozambique, and 6 South Africa. Luanda, SADC Energy Sector TAU.

recommended as a soil, water and draught-power conservation strategy[76] but perhaps due to lack of knowledge/education is not yet in use by the people of Xai-Xai and other rural areas across the country. However, because of the open grazing regime in Xai-Xai area, little crop residue remains in the field as cattle feed on the residue during the dry season. Furthermore, the roaming animals compact the soil thus rendering reduced tillage unattractive owing to poor water infiltration and consequently soil erosion. This might require that the people of Xai-Xai area be encouraged to develop a paddock system (reserved pastures) for their livestock. Below is a diagram for conservation farming that people of Xai-Xai could employ to successfully control soil erosion and maintain soil fertility:

Conservation farming
Adopted from D.J. BEUKES (FAO programme in Mozambique 2002-2006)

---

[76] For further discussion see FAO program in Mozambique 2002-2006, Maputo.

If conservation farming is carefully practiced especially alongside paddock system, it is likely to bring positive results on combating soil erosion and land degradation in Xai-Xai communal area.

To conclude this chapter, I underline that the risks faced by Mozambique due to climate change and environmental mismanagement related phenomena do not differ much from other countries located in sub-Saharan Africa. However, the high levels of poverty, illiteracy and, therefore, the capacity of the country to deal with the effects of climate change and environmental malpractices make Mozambique more vulnerable. Lack of infrastructures such as roads and shortage of qualified Agriculture Extension Officers to teach and empower peasant farmers in rural areas worsen the situation leading to prolonged challenges in health (spread of diseases), food insecurity, erosion and environmental degradation among other problems.

More importantly, I have argued that the problems and challenges encountered during the planning and implementation of environmental projects are not new or unique to Xai-Xai communal area, but are resonant of those encountered in other projects elsewhere in the country and beyond. Yet, the slow pace of transformation and environmental management skills transfer to communities hinder project community participation. This is what scholars like Abrams[77] protested against when he contends that in community-based projects the community should control the project and make important decisions, although professionals

[77] See report by Abrams, L.J. 1996. "Review of Status of Implementation Strategy for Statutory Water Committees", Department of Water Affairs and Forestry: Pretoria.

54

such as engineers may provide expertise and finance may be provided by external financial sources. In this regard, I have argued that for a community to control environmental projects, it must acquire administrative and environmental management skills. It has been made clear that although the planners' task is a crucial one, it should take into consideration the people at grassroots – those that are directly affected by the environmental problems in question.

In view of the results obtained from the study, it has been argued that although the Government of Mozambique has some national policies to enhance resilience of the poor populations in several vulnerable sectors, mainly in the area of disaster management, these policies are not directly tackling the issue of environmental and resource management problems. These are pieces of policies which are not implemented efficiently due to lack of expertise and coordination among the concerned sectors. There is need, therefore, to harmonize theory and action, that is, need for the harmonization of the existing policies, improving the information circulation and movement from emergency actions to preventive plans right from the grassroots levels.

Chapter 3

Green Revolution Program (GRP) in Mozambique: Rethinking the Impact of Mozambique's Fast-Track Green Revolution Program on the Environment and Biodiversity

The discourses on green revolution program (GRP) as the program has been exported to Mozambican rural areas are highly contentious and ironically present a maelstrom of technological and agricultural possibilities. Across Mozambique, the program has met with controversies of epic proportions, and indeed incited serious debates from various interest groups: academics, moralists, environmentalists and advocators of human and animal rights. While the program had the noble objective of attempting to reduce the country's food insecurity, its impact on non-human animals (heretofore referred to as animals), and the natural environment and biodiversity has resulted in earth-shaking transformations in terms of animal rights to life and unhindered access to space of habitation and damage to the natural environment. Since the 'call for the country to lead the green revolution (GR) in sub-Saharan Africa'[78] by the Republic of Mozambique's President, Armando Guebuza, there have been increased changes in the then status quo in terms of human and animal rights and balance of nature in the physical environment in

---

[78] For further discussion on green revolution in Mozambique see Africa News Network, 27/06/2008. "Mozambique green revolution will depend on small scale farmers", Mozambique Afrol News, http://www.afrol.com/Mozambique.BBCNEWS/http://news.bbc.co.uk/go/pr/fr//2/hi/africa/7601114.stm.

the countryside. The [negative] impact of the GRP in Mozambique has been largely a direct effect of uncontrolled veld fires which have intensified in the countryside since the start of the Mozambican green revolution program (MGRP). Veld fires are 'blazes that get out of control and devastate extensive tracts of forest, grassland, wildlife and other natural resources as well as injure and kill people and destroy their properties'[79]. They can be either natural or anthropogenic. Human beings are responsible for 95% of forest and veld fires,[80] as natural fires (fires not influenced directly by human beings) have become rare, especially in the sub-Saharan region.

The call for Mozambique to lead the GRP in sub-Saharan region before careful planning was done has resulted in hasty and unplanned form of agriculture by poor peasant farmers in the countryside. On the 27th of July, 2008, the President declared that 'the success of the green revolution lies in the hands of the family sector peasant farmers, and not of big projects such as Mozagrius'[81]. Mozagrius was a grandiose in the mid-1990s, whereby South African farmers were in the name of green revolution, to be attracted to Niassa province, where their mechanized farming would boost agricultural production. The declaration by the President had unprecedented effects on the countryside. About a month

---

[79] For further discussion on veld fires in southern Africa see Mkwanazi, H. 2007. Veld fire Campaign in Lupane. *Environment Africa*, Lupane.

[80] For further discussion on green revolution in Mozambique see Africa News Network, 27/06/2008. "Mozambique green revolution will depend on small scale farmers", Mozambique Afrol News, http://www.afrol.com/Mozambique.BBCNEWS/http://news.bbc.co.uk/go/pr/fr//2/hi/africa/7601114.stm.

[81] Ibid.

after the declaration, 'veld fires in the central provinces of Mozambique allegedly started by local farmers extending their fields, destroyed 16,000 hectares of arable land'[82]. According to the same source, many animals were displaced, killed and deprived of their habitats, and other such biodiversity were also destroyed, injured or have their habitats disturbed in some way. Animals and other biodiversity in the natural environment, thus, have been faced with new and sustained threats to their livelihoods and habitats as farmers venture into the havens of the helpless biodiversity with impunity for space to till. On the other hand, domestic animals particularly cattle and donkeys have suffered pitiful exploitation as the local farmers, poor as they are, have no farm machinery to use in their farms. The poor farmers' hope and success in farming lie in their domestic animals they use as draught power. Yet in the bid to maximize production, the farmers end up either abusing or 'unfairly' exploit the [domestic] animals in question. Makamure confirms this observation when he asserts: 'at the sorry end of human beings' excesses are the domesticated animals that have been used by humanity to till the land, as source of meat, clothes and ropes, to pull carts and other chores that make life for humanity better'[83] at the expense of suffering animals. Sad to note is the fact that the abuse and overexploitation of domestic animals though rarely discussed in conservation books, is not something new. In fact, the history of humans and animals (both wild and domestic animals) has always portrayed a morally disturbing inequality between the two species.

---

[82]Ibid.

[83] For a richer discussion on domestic animals in Africa see Makamure, D. M. 1970. "Cattle and Social Status" in Kileff C and Kileff. P (Eds), *Shona Customs*, Mambo Press, Gweru.

Humans have never accepted a situation whereby they put themselves on the same level with non-human animals moral-wise. Instead, a master-slave relationship has always existed especially between humans and the domesticated animals with the former considering the latter as resources to be exploited solely to improve the quality of human life.

Mozambican history, however, shows that before the chaotic fast-track GRP, animals, especially wild animal species had at least 'reasonable space'[84] of habitation; the right of animals to live in a natural free environment from human manipulation and interference was not severely threatened. The two, for instance, did not fiercely compete for space, neither did they compete for resources provided by nature for their sustenance and habitation. The relationship between humans and animals unfortunately turned otherwise with the advent of the so-called Mozambican green revolution program (MGRP). In the name of this program, land has increasingly become a scarce resource for both human beings and animals as competition for space between the two species gets stiffer year by year as humans keep on extending their farmlands. It is in this sense that one could argue that this unplanned fast-track program has overlooked the plight of biodiversity in the natural environment such as animals thereby bringing about unfortunate changes on the relationship between animals and humans with the physical environment.

As a result, the relationship gap between humans and animals has continued to widen in terms of moral and legal disparities. Although, both domesticated and wild animals throughout Mozambique's recorded history have suffered degradation, unimaginable and pitiful slavery, and possible

---

[84] Opcite

extinction at the hands of human race, their suffering has thus intensified since the onset of the fast track GRP in the country.

In the light of this observation, I content that the government, and in particular the ministries of Agriculture, National Farmers Union of Mozambique (UNAC), Wildlife and Forestry have all an uphill task to teach farmers good farming practices that do not upset the physical environment and violate other beings or at least animals' rights before continuing with their campaigns to call the country to lead the green revolution in sub-Saharan Africa. But what is this GRP we are talking about in view of Mozambique? This question calls for a careful unpacking of the concept of GR to enhance a better understanding of this study.

What is green revolution?

Since the beginning of agriculture on earth, people have been working to improve seed quality and variety. However, the term green revolution (GR) was never used. It was first used in 1968 by former USAID director, William Gaud to describe the transformation of agriculture in many developing nations that led to significant increases in agricultural production between the 1940s and 1960s[85]. This was partly triggered by the worst recorded food disaster known as the Bengal famine that occurred in 1943 in the then British–ruled India, killing four million people of hunger in eastern India. Talking of the Bengal famine, Amartya Sen[86] established that

---

[85] See Arundhati, R. 2004. "How deep shall we dig?" In *Asiatic Society*, Aligarh Muslim University Press, India.

[86] Sen, A. and Jean, D. K. 1989. *Hunger and public action*. Clarendon Press, Oxford.

while food shortage was a contributor to the problem, a more potent factor was the result of hysteria related to world war two, which made food supply a low priority for the British rulers. According to Sen, when it started, there were three basic elements in the method of the green revolution namely:

1. Continuing expansion of farming areas.

2. Double-cropping in the existing farmland (growing two crop varieties in two seasons per year, with water for second season coming from irrigation)

3. Using modern varieties – seeds with improved genetics (mainly wheat, rice, millet and corn).

The reason why these 'modern varieties' produced more than 'traditional varieties' was that they were more responsive to controlled irrigation and petrochemical fertilizers. With a big boost from the international agriculture research centres created by the Rockefeller and Ford foundations, the modern varieties also known as the "miracle seeds" quickly spread to Asia, and new strains of rice and corn were also developed. By the 1970s, the new seeds accompanied by chemical fertilizers, pesticides and irrigation had replaced the traditional farming practices of millions of farmers in developing countries and by the 1990s, almost 75% of the area under rice cultivation in Asia was growing these new varieties[87]. The same was true for almost half of the wheat planted in Africa and more than half in Latin America. In overall, a very large percentage of farmers in the developing world were using GRP seeds, with the greatest use found in Asia followed by Latin America.

---

[87] See Gaud, W.S. 1968. Speech to the Society for International Development.

The background, history and principles of Mozambican green revolution program (MGRP)

Mozambique's independence was achieved in 1975 after a ten year guerrilla war mainly by the *Frente de Libertação de Moçambique* (FRELIMO) against Portuguese colonial rule. For centuries, agriculture in Mozambique was characterized by subsistence farming, non-mechanized techniques and low yields yet the number of people in need of food has always been increasing. In realizing this, Mozambique made some attempts to move away from the so-called "primitive" farming techniques. One way out of this situation was to engage in the so-called green revolution program. 'Mozambique's attempt to launch a green revolution program dates back to as early as 1981'[88]. Malakata affirms this when he notes: 'the *new green revolution* has already been implemented in some African countries and even in our own country, without being named so'[89]. The move towards implementation of the GRP in Mozambique was prompted by the Ministry of Agriculture which admitted in 1981 that none of the state farms were profitable. Since then, several official explanations have emerged to account for such record of failure: excessively centralized management, poor control of stocks, insufficient infrastructure, poor use of machinery and lack of experience. In addition to these technical constraints, Hermele argues that social and political factors led to failure in the agriculture sector. In view of this, he

---

[88] Hermele, K. 1988. "Lands struggles and social differentiation in southern Mozambique: A case study of Chokwe, Limpopo", in Uppsala.

[89] For further discussion on Mozambique's green revolution campaign see Malakata, M. 2007. "Mozambique aims to lead green revolution": http://www.scidev.net/en/news/

notes that 'after independence, FRELIMO broke sharply with pre-independence policies in the liberated zones/rural areas'[90]. Instead of sustaining realistic alliances with some of the progressive, traditional, local authorities and existing social classes and market economy, FRELIMO pursued a policy aimed at the total transformation of Mozambican rural society based on wage labor in collective farms as was with the case of agro-industrial site at Chokwe in Gaza province of southern Mozambique. Unfortunately, the peasant farmers around Chokwe were not prepared to be agricultural labourers in state farm lands which they had originally occupied and then been expelled from first by the Portuguese colonial regime and later by the Mozambican government. For this reason, the farmers resisted the decision by government and consequently the progress in the state farms was very slow. Geffray[91] confirms this when he asserts that agricultural producer co-operatives suffered from labour, organizational and technical problems. The government's decision to embark on collective farming in state farm lands was premised on two major misconceptions: that they would develop spontaneously through mobilization of peasants, and that they would be rewarded by immediate increases in output. Unfortunately, neither of these assumptions practically materialized, yet the government's focus on large-scale agriculture had prevented it from thinking through the problems of expanding small holdings as a major theme in their economic program. Thus, the government's failure to provide support for peasant farmers at grass roots level resulted in some peasant farmers who had economic

---

[90] Opcite.

[91] Geffray, C. 1990. "La Cause das armes au Mozambique" Paris.

alternatives to abandon the state schemes and run their own "businesses".

The fourth and fifth party congresses of 1983 and 1989 respectively were an attempt to correct earlier mistakes and shortcomings, and heralding a new emphasis on more decentralized and capitalist-oriented small scale projects. It was also aimed at distributing land to peasant and private farmers, right from district level. However, state farms remained important, though they were to be oriented more towards production for export than to produce for urban markets in the country[92]. After the 16 year civil war that ended in 1992, the government remained focused on agriculture but now with incited intense competition over land resource, as many people who had fled out of the country during the war had come back. Besides, soaring prices, elimination of nearly all food subsidies and high levels of unemployment ignited by Economic Structural Adjustment Program (ESAP) in the early to mid-90s resulted in most families, both in the rural and urban areas, realizing the critical importance of having a piece of land to grow basic crops for their families and sell the surplus. Competition on land thus increased immensely.

Though it is true that Mozambique's agricultural sector was growing very slowly, Mozambique had to use agriculture as one of the strategies to get out of her economic squalors. In the mid-1990s, Mozagrius, a grandiose scheme of South African farmers was attracted and invited by the Mozambican government to Niassa and Tete provinces to bring their mechanized agriculture to these areas so as boost agricultural

---

[92] Roesch, O. 1989. "Economic reform in Mozambique: notes on stabilization, war, and class formation" in Taamuli, Dar es Salaam.

production in the country[93]. But though, some of the South African media compared the scheme to a second "great trek,"[94] only a dozen or so South Africans came to Niassa and most of these left within a few years. The profits they expected were never forthcoming and the scheme collapsed. This is one other reason why Rodman and Gatu[95] conclude that neither the state, market, farmers nor the geopolitical context are working in favour of a GRP in Mozambique, as the farmers have to deal with too many obstacles if a green revolution is to be possible. In the 2004 and 2009 Presidential campaign manifestos, Armando Guebuza promised that he would promote agriculture and do away with corruption in order to eradicate poverty in Mozambique. In 2007, 'President Guebuza calls Mozambique to aim at leading green revolution in sub-Saharan Africa'[96]. As one way of calling the rural population to focus on agriculture, the President further pointed out that 'large scale projects may not respond to the goals of the green revolution because their primary objective

---

[93] UN Report, 2008. "The food and Agricultural Organization of the United Nations", The Nutrition Country Profile. Werichannel news Available at: http://werichannel.wordpress.com/2008/09/18 scores die-inmozambican- veld-fires/

[94] Ibid.

[95] Rodman, S. and Gatu, K. 2008. "A Green Revolution in Southern Niassa, Mozambique?: A field Study from a small Farmer Perspective about Possibilities and Obstacles for a Green Revolution", Växjö Universitet.

[96] For more on green revolution in Mozambique see Malakata, M. 2007. "Mozambique aims to lead green revolution"; Also available online at: http://www.scidev.net/en/news/

is to make profit'[97]. He, however, acknowledged the role played by commercial farmers in increasing agricultural production at national level, but reiterating that the family sector is the key to success in food self-sufficiency in the country. This calling for a green revolution by the Mozambican president has, however, resulted in more problems created than solved, especially between humans and other biodiversity such as animals. In the ensuing sections, I highlight problems associated with the Mozambican GRP before discussing the [Western] conception of the relationship between humans and animals throughout recorded history and demonstrate how this relationship has aggravated in the context of the MGRP. This is done in attempt to generate knowledge in African conservation literature that looks at the classical and current conceptions of the relationship between humans and animals.

The GRP and its problems in Mozambique

Green revolution technology in Mozambique like elsewhere in the world is aimed at increasing agricultural production and reduction of the country's food insecurity. However, in Mozambique, the GRP has created more problems than it intended to solve, not only to the physical environment, but also to biodiversity including nonhuman animals and humans themselves. For example, competition for land resource between humans and animals has become fiercer than ever. As espoused by Moyana, land can be

---

[97]See Africa News Network, 27/06/2008. "Mozambique green revolution will depend on small scale farmers", Mozambique Afrol News, http://www.afrol.com/Mozambique.BBCNEWS/http://news.bbc.co.uk/go/pr/fr//2/hi/africa/7601114.stm.

understood as a natural resource that is possibly used as an instrument of production to give life and means of survival to both humans and animals. This definition implies that animals just like human beings are entitled to rights for life, free living space and interaction with their environment and even with each other. In Mozambique, this understanding however, seems to be too far from being realized in the face of GRP as the relationship between humans and animals with the physical environment is ever aggravating.

From the Western classical view of human-animal relationship to the human-animal relationship perspective in Mozambique

Though there is dearth of Mozambican or rather African literature on the relationships between humans and nonhumans such as animals, it appears that throughout Mozambique's recorded history, the human race has tended to view animals (both wild and domestic) as nothing of any moral status, but natural resources for their own good/ends. Plato (428-348) set a strong philosophical tradition on the question of the relationship between the human race and animals. He regarded man as primarily not an animal at all, but a superior being with rational powers to dominate other entities in the realm of existence such as animals and plants[98]. Plato's characterization suggests that man being rational is an animal of a higher order that deserves moral consideration and treatment completely different from that of animals. In his theory of evolution, Plato affirms that all other animal species have descended from man. He notes 'the descent is downward and the first step from man is woman and

---

[98] Miller, H. B. 1983. "Platonists" and "Aristotelians" in Miller. H. B. and Williams. W. H (Eds) *Ethics and Animals,* Humana Press, Clifton.

children….animals occupy the lowest rank on the moral scale'[99]. In view of this understanding, Plato argues that animals cannot expect to be accorded moral status equal to that of man or at least any moral status.

Aristotle (384-322) seconded his teacher's position in view of the relationship between animals and the human race when he argues that animals should never be accorded any moral status for the reason that they are irrational. Aristotle identifies three levels of life. These are nutritional/vegetative (plant life with powers of reproduction), animal (with powers of sensation including those of the lower levels) and human lives (intellectual powers including those of the lower levels). Since human life is at the highest level, it follows that 'there is more to life in man than in a dog, in a dog than in a worm, in a worm than in a plant, and in a plant than in a stone,'[100] hence, Aristotle's famous *Scala naturae* – ladder of nature. From the understanding that life is characterized by the "psyche" which thinks, reasons and wills; life has been technically construed as 'the condition that distinguishes animate from inanimate things including the capacity for growth, reproduction, functional activity and continual change preceding death'[101]. The point has to be made, however, that for Aristotle, since vegetative life is inferior to animals and human life, vegetative life must meet the needs of animals and human beings. Likewise, because human beings are endowed with the faculty of reason, they are

---

[99] Ibid.pp.2.

[100] O'Connor, D.J. 1985. (Ed). *A Critical History of Western Africa*, Free Press, New York, pp. 53.

[101] Ibid.

superior to animals. The primary function of animals is, therefore, to serve the needs of human beings[102].

Aristotle further defended his view using a doctrine of natural slavery in which he ranks in a decreasing order of merit: man, woman, child, natural slave and lastly animals. In his words, 'the ox is the poor man's slave'[103]. This shows that animals are viewed in a degrading way in the Aristotelian world view. Thomas Aquinas (1225-1274) agreed with Plato and Aristotle that human beings are more superior to nonhuman animals. He ruled out the 'possibility of sinning against animals or the natural world,'[104] if humans exploit them for their benefit. Aquinas identifies the faculty of reason as the central feature that distinguishes human beings from animals. Following Plato and Aristotle, Aquinas argues that only rational beings are capable of examining and determining their actions. Other beings like animals cannot direct their own actions. For this reason, only human beings can be ascribed intrinsic moral worth. Animals only have instrumental value and not intrinsic moral worth as they exist in order to be used in different ways by human beings.

Immanuel Kant (1724-1804) holds a similar view. For him, 'rationality and autonomy are the central features that only human beings and not animals have'[105]. Though, Kant acknowledges that both human beings and animals have desires that urge them to perform certain actions, only human

---

[102] Miller, H. B. 1983. 'Platonists' and 'Aristotelians' in Miller. H. B. and Williams. W. H (Eds) *Ethics and Animals,* Humana Press, Clifton.
[103] Ibid.pp.2.

[104] For further discussion on animal and environmental ethics see Singer, P. 1993. *Practical Ethics*, Cambridge: Cambridge University Press.

[105] Boss, J. A. 1999. *Analysing Moral Issues*, Mayfield Publishing Company, Belmont, pp. 25.

beings can reflect on their actions and determine whether the actions are worth performing or not. Out of this reflection, human beings can act out of good will and reason. Other animals cannot do this as they lack rationality and autonomy.

Boss, a contemporary scholar, has also argued against giving moral rights to animals. For him, 'animals lack the capacity for autonomous moral judgment and reasoning'[106]. It is out of this understanding that Boss reasons that only human beings and not any other animal deserve direct moral consideration.

Now coming on to the humans-nonhuman animals relationship in the face of Mozambique's GRP, it is unfortunate on the part of animals to note that the general conception of the above scholars have been transported into the present day Mozambique to the extent that it dramatically influence the whole country's societies. Speaking historically of Mozambique, one would learn that domestic animals such as cattle and donkeys have always been used to work in the fields and for draught power. Wild animals as is the case with most of the domestic animals, have also been used as sources of meat, leather, ropes and milk. The Mozambican societies especially the contemporary ones involved in the GRP share the views of Western scholars like Plato, Aristotle, Aquinas, Kant and Boss that nonhuman animals are merely resources to be exploited to advance human needs and interests. From the findings of the researches I carried out in many parts of rural Mozambique, it was evident that the farmers involved in the MGRP are not ranking animals on the "moral ladder" as they think animals since the beginning of history were meant to serve and satisfy the interests and needs of human beings. While Mozambican societies acknowledge that both wild and

---

[106] Ibid.

71

domestic animals have traditionally played a pivotal role in the life and history of the human race in serving as sources of meat, milk, clothing, draught power and in various traditional ceremonies, they think animals have always had instrumental rather than moral value. The call by the president of the Republic of Mozambique to lead green revolution in sub-Saharan Africa has resulted in the intense use (as means to their ends) of cattle and donkeys by poor farmers in the rural provinces of Manica, Sofala, Tete, Nampula, Niassa, Zambezia and Cabo Delgado who are occupying the new farming areas but without enough or proper farm machinery to cultivate the land. According to Allafrica,[107] most of the people in Majune district (in Niassa province of north Mozambique), one of the districts with great potentials for agricultural production, live in abject poverty. Domesticated animals such as cattle and donkeys are suffering the consequences as they are the sources of labour in the fields of the people who live in the area. Worse still, these animals are denied the right to share the proceeds (from the fields) after harvest. For example, if caught in the field which the oxen or the donkey itself helped in cultivating, the animal is ruthlessly beaten up. Other domesticated animals like dogs suffer the same moral inequalities from the human race. Because the farmers' fields are not fenced, dogs are serving as field guards to potential animal pests, which might destroy crops. Unfortunately, the dogs receive nothing in the end, not even the right to decent meals. This morally disturbing inequality between animals and the human race is a clear testimony that

---

[107] Africa News Network, 27/06/2008. "Mozambique green revolution will depend on small scale farmers", Mozambique Afrol News, http://www.afrol.com/Mozambique.BBCNEWS/http://news.bbc.co.uk/go/pr/fr//2/hi/africa/7601114.stm.

Mozambican contemporary societies share the Western traditional view of man's dominion over animals and the latter's use as means to human ends. Such a pitiful relationship between animals and human beings can be equated to a master-slave relationship whereby a slave lives only to serve the interests and needs of his [her] master.

It is from this observation that my conception of the relationship between humans and animals identifies with Jeremy Bentham (1748-1832) and his follower J.S. Mill (1806-1873) who advocate the ascription of moral status to animals. Bentham, the founder of hedonistic utilitarianism held that, an act is right if it brings about the greatest net amount of pleasure, wrong if it brings about net sadness to those affected by the action. Bentham for instance, had it that the important feature to put animals in the moral realm must not be the reason or ability to talk, but the fact that they can suffer[108] in much the same way as human beings.

In the like manner, Regan vehemently argues against the use of animals in factory farms, laboratories and zoos since under such conditions the animals are used as mere means to human ends; yet, 'if human beings have rights so are non-human animals'[109]. Animals thus are morally equal to human beings. Thus for Bentham, Regan and rightly so, the "unfair" use of donkeys, dogs and cattle by those involved in the MGRP would have not only incited serious moral questions, but also set an uphill challenge to the Mozambican president and his government who called for peasant farmers to engage in an unplanned GRP.

---

[108] For further discussion on animal ethics see Raghavan, S.M. 1999. Animal Liberation and "Ahims" in Boss. J.A.(Ed) *Analysing Moral Issues,* Mayfield Publishing Company, Belmont.
[109] Ibid.

In view of what is transpiring in the newly settled areas between human and animals, Peter Singer rightly advocates for animal liberation based on Bentham's utilitarian theory. Singer thus would remark: 'sentience and not reason is the criterion that ought to be used to consider a being's moral status'[110]. Sentience is the capacity to feel pain and pleasure. In this light, since the cited animals are working but receiving nothing in the end, they, just like human beings feel pain. They should, therefore, be accorded moral consideration equal to that of human beings. Thus for Singer and rightly so, no objective assessment can support the view that it is always worse to oppress members of our species who are persons than species who are not given that nonhuman animals are companions of humans in the environment they share.

From the foregoing, one would accuse those who deny animals' entitlement to moral consideration for committing a fallacy called "speciesism." Denying other sentient beings right to moral consideration is denying the same rights to oneself. Wild animals have not been spared by the green revolution farmers' cowardice. In pre-colonial Mozambique, 'the right of wild animals to live in a natural environment free from human manipulation was not yet severely threatened,'[111] since human population was still very low. The human race and wild animals did not fiercely compete for space and other resources for survival. The advent of the Portuguese settlers and later the GRP in the recent years, have however reversed the situation. The most important natural resource – land – has since then increasingly become a scarce resource, both to

---

[110] For further discussion on ethics see Singer, P. 1993. *Practical Ethics*, Cambridge: Cambridge University Press.

[111] Wolmer, W. 2007. *From Wilderness Vision to Farm Invasions: Conservation and Development in Zimbabwe's South-East Lowveld*, Weaver Press, Harare.

74

the Mozambicans and the wild animals. Wild animals have been unfortunately deprived of their rights to land resource and food products from nature as the new farmers venture into their havens with impunity for meat and space to till and settle. Thus, throughout recorded history, the relationship between animals (both domesticated and wild animals) and the human race have always shown serious moral disparities that should be renegotiated.

The life of the human race thus, has always been improved at the expense of suffering animals. Mozambique's disorganized so-called green revolution program seems to have overlooked the plight of animals as new farmers scramble to grab and clear land that used to be havens for the animals. Most of the wild animals have been left stranded thereby falling prey to the new farmers themselves. Some of the species existent such as *mavhondo* (rabbit-like animals), hares and elephants which used to exist in very large numbers in the rural areas of Zambezia, Manica, Sofala and Tete provinces have since been threatened[112].

It is for this reason that I identify with Schmidt Raghavan[113] who argued for the moral status of animals, of course not as equal beings to humans, but at least to the extent that they have the right to live freely in their natural habitats. For Raghavan, all living things including animals have inherent moral value and ought to be treated with equal respect. She remarks 'human beings ought to refrain from

---

[112] Africa News Network, 27/06/2008. "Mozambique green revolution will depend on small scale farmers", Mozambique Afrol News,http://www.afrol.com/Mozambique.BBCNEWS/http://news.bbc .co.uk/go/pr/fr//2/hi/africa/7601114.stm.

[113] Raghavan, S. M. 1999. Animal Liberation and "Ahims" in Boss. J. A.(Ed) *Analysing Moral Issues,* Mayfield Publishing Company, Belmont.

killing and causing pain and suffering to helpless animals when alternative and more economical sources are available'[114]. In view of Raghavan's argument and in the light of what is happening in Mozambique's GRP, the suffering and mass onslaughts which these animals are facing is morally unjust and unacceptable. Yet the GRP in Mozambique has not only [negatively] impacted on the lives of nonhuman animals but the physical environment in general.

The impact of MGRP on the physical environment

Traditionally, Mozambicans have an environmental ethic that considers the interests and needs of the whole natural environment as demonstrated by their use of philosophies of taboos, totemism, folktale and ubuntu. I will not elaborate on each of these philosophies in terms of their respect for the environment as I have done this elsewhere (Mawere 2011; 2012; 2013). For now, it suffices to say that all these philosophies were meant to guarantee a harmonious, perpetual sustainable interaction between humans and all the constituents in the environment they shared. This is to say that traditionally, Mozambicans are totally against wanton destruction of both fauna and flora. Anyone who wantonly destructs vegetation was considered unethical and inhumane. The culprit would be arrested by the chief's policemen and tried by the chief's court. In Mozambique and by extension Africa, the natural environment is said to contain sacred places which are residing palaces of the ancestral spirits – *mhondoro* (lion spirits) and *njuzu* (mermaids). Such places include rivers, mountains, forests and sacred curves.

---

[114] Ibid.

With the advent of the Portuguese settlers and later the invasion of land by local peasant farmers under the so-called MGRP, a negative relationship has developed between the Mozambicans and their physical environment. Sacred places like mountains and forests have suffered immense destruction from human induced fires and other such anthropogenic activities. Afrol News[115] reveals that more than 16,000 hectares of arable land in three central provinces of Mozambique (Manica, Sofala and Nampula) were recently destroyed by veld fires allegedly caused by farmers extending their farm lands. These fires contribute to a significant proportion of land degradation and greenhouse gases to the atmosphere.

Furthermore, veld fires destroy resources needed for immediate use over the dry season, crops, firewood, water sources and grazing land. Frequent veld fires may also have long-term effect on the reproductive capacity of important veld products such as vegetation and humus. Even more troublesome, the culprits are rarely identified, held accountable or reprimanded. Besides, many incidences of the damage to the physical environment by such fires go unreported as some rural areas are too remote for news reporters to access easily. The director of Mozambique's relief agency, Joao Ribeiro confirms this when he acknowledged: 'this only reflects to areas where fires are known and the number of victims might be higher than officially recorded'[116].

---

[115] See Africa News Network, 27/06/2008. 'Mozambique green revolution will depend on small scale farmers', Mozambique Afrol News,http://www.afrol.com/Mozambique.BBCNEWS/http://news.bbc .co.uk/go/pr/fr/-/2/hi/africa/7601114.stm.
[116] See UN Report, 2008. "The food and Agricultural Organization of the United Nations", The Nutrition Country Profile. Werichannel news

In a recent study, *Allafrica*[117] observes that farming in Majune district in Niassa could wipe away the scenic beauty of the area even before it can realize its potential as a successful ecotourism product, as evidence of indiscriminate burning of the bush are common and nobody seems to care. Illegal settlement by farmers trying to fight poverty by exploiting the natural resources particularly the forests is a common phenomenon. The forest is usually burnt wantonly to create space for a new farmland and settlements whose residents embark on charcoal production and falling poles for sell. All these activities degrade the physical environment and upset the balance of nature. Thus, extensive education and sound agronomic practices are necessary as they can markedly help in minimizing environmental damage and increase crop productivity and yield. By default, the failure of the state to provide appropriate legislative interventions and to educate the new farmers under the green revolution program good farming practices that does not violate human rights makes the state an accomplice in this arrangement. It is the contention of this chapter and indeed the whole book, therefore, that the government, and in particular the ministries of Agriculture, Wildlife and Forestry have an uphill task to teach farmers good farming or agronomic practices that does not upset the physical environment and violate both humans and nonhuman animals' rights, before continuing with their campaigns to call the country to lead the green revolution in Africa.

To conclude this chapter, I underscore that the history of the relationship of human beings and biodiversity, and in particular animals and the physical environment has been

---

http://werichannel.wordpress.com/2008/09/18 scores die-inmozambican- veld-fires/

[117] Opcite.

examined. I have argued that throughout recorded history, the relationship between human beings and the domesticated animals have been that of the master-slave. That of human beings and the wild animals has always been likened to the relationship between cats and rats, where the former always hunt the later. Yet in Mozambique, this relationship has worsened with the establishment of GRP. In fact, the chapter has demonstrated that in Mozambique, the relationship between the human race and biodiversity such as animals has been aggravated by the fast-track and pre-planned green revolution program. Through the program, the physical environment has not been spared from fatal degradation as the balance of nature has been dramatically threatened and disturbed especially by anthropogenic fires and other such activities. In view of these observations, I have argued on the basis of life ethics and environmental ethics for the need to rethink relationships, interactions and networks between humans and entities in the natural environment. Serious consideration of the moral status and rights of the physical environment and non-human animals, thus, has been stressed.

Chapter 4

Gold Panning In Central Mozambique: A Critical Investigation Of The Effects Of Gold Panning In Manica Province

Gold panning, also known as artisanal mining is a poverty driven occupation present in over 70 countries[118]. For purposes of this work, the term gold panning shall be used more often than artisanal mining. Gold panning can be legal or illegal. In some cases especially where there are associations, gold panning can be organized, as medium or large-scale but in most cases it is a disorganized and small-scale operation. Also, though men other than women are the societal members mostly involved in the activity, gold panning is not a male gendered operation as it directly or indirectly involves women and children alike. In many countries, gold panning lacks government support resulting in lack of education by those involved. This explains one of the reasons why processing techniques used by gold panners in the extraction of gold vary ranging from rudimentary or manual panning to semi-mechanized. Besides, lack of education especially among miners in the developing world explains why gold panning is normally associated with human health and environmental related problems as well as the misuse of hazardous materials (used during extraction or processing of the mineral) such as mercury and cyanide.

---

[118] See Interim report of the UNEP Global Mercury Partnership by Telmer, K. and Veiga, M. M. 2008. World emissions of mercury from artisanal and small scale gold mining. In: Pirrone, N. and Mason, R. (Eds). *Mercury fate and transport in the global atmosphere: measurements, models and policy implications.*

That being the case, Mozambique like other developing nations the world over is at the risk of facing catastrophic changes in the physical environment, water sanitation, agriculture, biodiversity and habitats and, of course, new pressures on human healthy and economy. These risks are a result of a myriad of factors ranging from environmental mismanagement, poor exploitation of resources, poor agriculture, climate change and gold panning. In Mozambique, and in particular Manica province, poor exploitation of resources such as gold and poor agriculture seems to be among the top factors putting the province at risk. The risks are generally manifested in pollution, poor human health, environmental degradation and dwindling of the agriculture sector. Such risks threaten to worsen the existing levels of poverty, food insecurity and undermining all national development efforts. It is in view of this realization that I advance the thesis that in Mozambique, since about 80% of the population depends directly on land and natural resources from the natural environment, the effects of climate change, gold mining, environmental damage, water pollution and variability are likely to have a great influence on the communities and in the national economy in general. According to Mozambique's initial Communication to UNFCCC,[119] the main sectors likely to be impacted by both natural and anthropogenic causes and climate change include: agriculture and food security, water resources, costal resources, biodiversity, human health and infrastructure, loss

---

[119] For further discussion on climate change related issues in Mozambique see Mozambique First National Communication to the United Nations Framework Convention on Climate Change (UNFCCC), 2006, Submitted to the Secretariat of the Convention in 2006. Maputo – Mozambique; See also Bambaige, A. 2008. National Adaptation Strategies to Climate Change Impacts: A Case Study of Mozambique. World Human Development Report Office.

of life, erosion, land degradation, sea level rise, natural disasters, salt intrusion, crops, ecosystems, property, human and animal habitats, outbreaks of pests and diseases, displacement of people, and destruction of infrastructure (communication network, schools, hospitals and houses). This situation is aggravated by the fact that Mozambique is home to many gold panners estimated to be at least 60,000 artisanal gold miners,[120] 18,000 of whom are women and children. The artisanal gold miners produce over 90% of the country's gold[121]. It is worth noting that majority of these artisanal gold miners are concentrated in the Manica province in central Mozambique where some come from the neighbouring countries such as Zimbabwe, Malawi, South Africa, Botswana and Zambia. According to recent studies, 'in Manica province more than 20,000 fortune hunters are digging for gold in the mountains of Mozambique. Scores have come across the border from bitterly poor Zimbabwe. But for most, the dream of fabulous treasure ends up in endless toil, disease and death'[122]. Such great numbers of gold hunters are disturbing as they put the province at an unimaginable socio-economic risks and environmental threats especially considering that gold panning in this part of the country is largely uncontrolled.

---

[120] See MMSD, 2002. Breaking new ground: mining, minerals and sustainable development, *International Institute for Environment and Development*, London.

[121] Mondlande, D.S. et al. 2002. The Socio-economic impacts of artisanal and small scale mining in the developing countries, *Blackwell Publishers*, Rotterdam, The Netherlands.

[122] See Spielgel International, 24/03/2008. The gold slaves of Mozambique, Available online at: *http://www.spielgel.de/international/topic/africa/*

While all the above mentioned problems are associated with gold mining in Mozambique, what remains worrisome when it comes to researches on gold mining in the country is the fact that most of them[123] draw more attention on ways of reducing mercury pollution (used in gold amalgamation) without addressing other 'serious' problems being caused by illegal gold panning such as environmental degradation and pollution. The history of mining in Mozambique thus makes a sorry reading with its failure to document, by default or otherwise, some detailed accounts of other serious effects of gold panning in the country, especially in the gold affluent Manica province. Following this observation, the thesis of this chapter is that gold panning, though has become a major source of livelihood for many people in Manica province and beyond, is causing more harm than good to humans, non-humans and the natural environment at large. Recommendations on what needs to be done on the ground and conclusions are drawn towards the end of the chapter.

---

[123] For further discussion on gold mining in Mozambique see Zacarias, R. & Manuel, I. 2003. Assessment of mercury use in artisanal gold mining in the mining in the Manica district of Mozambique, In: Artisanal and small-scale mining in developing countries, *Urban health and development bulletin*, Vol. 6 (4) 57-61; See also Spiegel, S.J. et al., 2006. Mercury reduction in Munhena, Mozambique: homemade solutions and the social context for change, International Journal of Occupational and Environmental Health, *Scopus*, 15, p. 215–221; See also Shandro, J.A. et al., 2009. Reducing mercury pollution from artisanal gold mining in Munhena, Mozambique, *Journal of Cleaner Production,* Vol 17 (5) 525-532; See also Blacksmith Institute Report, 2005. Pilot project for the reduction of mercury contamination resulting from artisanal gold mining fields in the Manica district of Mozambique, *Global Mercury Project*, Austria; See also Blacksmith Institute Report, 2011. Gold mining and mercury emissions in Manica, Mozambique. Available @ *http://www.blacksmithinstitute.org/*

Background and historical overview of gold panning in Manica

Mozambique like other countries in sub-Saharan Africa has a long tradition of mining, and in particular gold mining. The mining activity has always been associated with serious problems especially where there are no associations or at least government control or interventions. As such, gold panning is not a new and unique problem to Mozambique, but is ubiquitous in other countries in the region and beyond.

In Mozambique, gold mining in general dates back to the pre-colonial era, that is, to the time before the advent of the Portuguese settlers. Historical data reveal that gold in Mozambique was mined by Zimbabwean states between 1250 and 1450 AD, and the Mwenemutapas between 1325 and 1600AD. During the colonial era, some large international companies were authorized by the Portuguese colonial government to mine the precious mineral, gold. However, in 1888, discoveries of gold were made in Witwatersrand in South Africa, causing a "gold rush" in Manica. Many of the European companies migrated to South Africa. This is confirmed by Manuel *et al*[124] who note that due to the discovery of gold in Witwatersrand, by year 1900 twenty-three gold mining companies had left the province. According to the same source, between 1900 and 1949, about 9,530 kgs of gold were extracted in Manica province 63% of which was from alluvial deposits. Due to intensity of the war of liberation in the country, between 1949 and 1975, the production of gold declined significantly. This situation continued into the 1980s as this period was a destabilization

---

[124] See Manuel, I.R.V. et al. 1999. Exploração artisanal do ouro no distrito de Manica: Degradação ambiental versus desenvolvimento, *Congresso Luso-Mozambique de Engenharia,* Maputo.

phase characterized by internal instability accentuated by civil war. As a matter of consequence, gold mining in the province practically came to a standstill during the period between 1975 and the last years of the 1980s.

With the return of peace in 1992, mining across the country especially in the conflict-torn Manica province resumed. The *Aluviões de Manica* (ALMA) company, a joint venture between the Mozambican government and the British's LONHRO, began gold exploration in Ruvue River basin in 1990 and started operations immediate after the civil war. In many other parts of the province, gold panning heightened between 1992 and 2000 due to erratic rainfalls and frequent severe drought that hit across the region. Gold panning, thus, became the most and sole source of livelihood for the people living in the region. In Munhena mining site in Manica, for example, gold mining started in the pre-colonial era but only resumed in 1999 with the discovery of a gold deposit by a young girl. Other gold deposits scattered in the province like those in Chazuka, the Siyawatonga Mountains and Mutowe areas also have a pre-colonial history, but mining resumed after the civil war and is continuing even to date.

Though artisanal gold mining in Manica is largely uncontrolled by the government, currently there are three big gold mining companies in the province namely AUSMOZ, JF Mining, and *Exploracões Mineiros de Mozambique*. While mining by these companies is organized and that at Munhena mining community a bit more or less the same as it comprises an association of over 3000 members who work on a 25 year lease, mining in many other sites across the province is uncontrolled. Government granted 143 ha concession and generates a substantial income (producing over 5 kg of gold per month) from the aforementioned companies but receives no revenue from the smaller mining sites across the region where artisanal mining is uncontrolled.

Given that the biggest activity in Manica province is artisanal mining (both legal and illegal mining), mining remains the biggest source of water pollution and environmental degradation in the province. Researchers have shown that there are high mercury levels in the rivers flowing through Manica, besides sediment concentrations that any sight gifted person can observe. This is aggravated by illegal gold mining across the province that remains disorganized and continues to engage in activities that are perilous and indeed a big threat to the well-being of humans, non-humans and the physical environments. Shandro *et al*,[125] for example, noted that gold panning associated environmental and health costs are high in Manica as mercury continues to be used and lost to the environment, and cyanide will be introduced soon. This is worrisome as almost all miners in Manica continue using mercury to amalgamate gold. Mercury is a heavy toxic metal with potential to bio-accumulate in the food chain. It is hazardous to aquatic environments (fauna and flora) and humans (miners and people downstream) and intakes can occur via food, water and air. Mercury exposure amongst workers and community members associated with artisanal gold mining is therefore high, and has been well documented[126]. This makes gold panning a perpetual potential threat to all forms of life in the province though gold is one of the major sources of livelihood for many

---

[125] See Shandro, J.A. et al., 2009. Reducing mercury pollution from artisanal gold mining in Munhena, Mozambique, *Journal of Cleaner Production,* Vol 17 (5) 525-532.

[126] See Swain, E. B. et al., 2007. Socio-economic consequences of mercury use and pollution, *Ambio* 36, pp. 46–61.

people in the region. Of concern is that though there are some strides by the government to address the concerns arising from mining in the province, the action is too slow for this mounting and fast spreading problem.

Having shown the background of the problem studied, the next section takes a look at research questions and the methodology used in carrying out the study.

Research questions and methodological issues

The present study seeks to address the following questions:

1) What are the effects of gold panning (to the physical environment, humans and nonhuman animals) in Manica province besides mercury pollution?

2) What measures could be put in place in order to address the problems in the mineral affluence Manica province?

As part of research design for my study, I relied on literature studies, content analysis, field observations and in-depth interviews. The research was carried out between June and July 2011 using a randomly selected sample of 26 people (13 female and 13 male). The sample size of 26 was considered sufficient in providing the general perceptions of the people of Manica province, particularly the directly affected individuals.

Using field observation data collection procedure, I observed the physical environment and the environmental malpractices in the chosen area. The field observation, a method which I adopted from my anthropological studies, was used as one of the major collection tools to ascertain the project location and what really happened on the ground. Observation allows the researcher to have access to first-hand information that he [she] can observe and record in person. As such, I personally made the observations around the mining sites and the affected environments such as rivers, plants and land.

To supplement the field observation data, informal interviews were conducted, particularly with the mostly affected (directly or otherwise) members of the society. This was done in order to obtain more information on the possible effects of gold panning in addition to those I got through observation. More so, I wanted to hear from the affected people – 'the real affected communities' – on what they think could be lasting solutions to the problems they are encountering. This is important as affected people understand their problems better than anyone else.

The people participated in the study were from different societal classes, ranging from gold panners, farmers, fishermen, students and religious people (both Christians and traditionalists). The respondents were drawn from different societal classes with the hope for obtaining a balanced research result that could be representative of the whole affected areas. They ranged from 12 to 50 years. This age group was considered appropriate for the study because it is within the range of the active group mostly involved in societal activities to do with natural resources exploitation and environmental management. Equal number of men and women were sampled for the mere reason that both sexes seem to be equally involved, directly or otherwise, in gold panning. The participants interviewed responded to the questions individually and participation was voluntary. Participants were assured of the confidentiality of their responses and were asked not to identify themselves by names or otherwise. The collected data were tabulated to show frequencies before being subjected to evaluative critical analysis. The Tables 3 and 4 respectively contain details of the people participated in the study and the data that were gathered during the study:

Table 3: Participant demographics

Occupation	Gender	
	Male	Female
Farmer	3	3
Gold panners/miners	2	3
Fisherman	2	2
Religious people	2	3
Students	2	2
Village head	2	0

Source: *Survey 2011*

Research results
Table 4: Responses to closed questionnaire items

ITEM	RESPONSES		
	Agree	Disagree	Uncertain
1.Gold panning is making water unsafe for drinking	26	0	0
2.Gold panning is destroying aquatic flora and fauna	24	0	2
3.Prostitution and violent related problems are increasing in gold mining sites	19	1	6
4. Irrigation and fishing in rivers near gold mines have been negatively affected	22	1	3
5. Disease outbreaks are more prevalent in mining sites than elsewhere in countryside	23	2	1
6. Illegal gold panning is robbing the state of its revenues	14	2	10
7. Gold panning in Manica is only benefiting the minority	20	2	4
8. Gold panning should be stopped in Manica	21	5	0
9. Gold panning should be government controlled i.e. done by licensed people	23	3	0
10.Government and humanitarian organizations should offer health and education programs/services to people at mining sites	23	3	0

Source: *Survey 2011*

Discussion based on field observations and interviews

The findings that are presented in this research are based on the data that were collected from the "natives" or people who live in Manica province between June and July 2011. They show both positive and negative perceptions on gold panning. This means there were mixed feelings with regard to the gold mining activity, particularly gold panning in the province.

The notion that gold panning was making water unsafe for drinking was uncontested as all respondents (100%) agreed. Possibly, the reason for this was that reddish-brown sediments are visible all over the affected rivers such as Lucitu and Musapa Rivers, which makes it difficult to deny the fact that gold panning is polluting the waters. Almost similar results (92.3%) were obtained on whether gold panning is destroying aquatic and fauna biodiversity. One of the respondents revealed that a number of dead fish have been witnessed along Lucite river since the previous year, 2010, with the discovery and intensification of illegal gold panning upstream in the Musanditevera mountains in Mutowe area of Manica province. More or less the same results (84. 6%) were also obtained on questions whether gold panning is hindering agriculture, fishing and accentuating prostitution, disease outbreaks and cases of violence in the area. Reasons given varied but the major one was that gold panning in Manica is largely uncontrolled by the government; the "law of the jungle" prevails in most of the mining sites. This suggests the government should put up measures to control the mining activity in the province.

There was lack of knowledge by some respondents on whether illegal gold mining results in loss of government revenue as 38. 4% of the respondents were uncertain if illegal gold panning robes the state of its revenue. It appeared that most of these respondents didn't know what government revenues are. I took my time and patience to explain the term to participants. This was done to ensure that responses from informed positions were obtained.

On whether gold panning should be stopped, an overwhelming majority (80.7%) of the respondents disagreed, thus showing support for the mining activity, though in most cases done illegally and its results a threat to both human and the natural environment's well-being. This showed that gold panning as a means of survival by the locals is so deeply woven into the fabric of traditional socio-economic life in Manica that the people have accepted it and even unconsciously. However, most of the respondents (88. 4%) were quick to point out that though gold panning is one of their major means of survival, it should be government controlled and carried out in an environmentally safe and friendly manner. This finding concurs with results from a recent study on Mozambique's green revolution program I carried out elsewhere (see this volume) which notes that if no action is urgently taken by the government to control anthropogenic impacts on the natural environment there will be increase in natural hazards, mortality rate, mean air temperature, reduction of annual rainfall and increase in solar radiation. In the light of such findings and projections, I make a clarion call for the Mozambican government to reconsider illegal gold panning in Manica province, to incorporate an environmental ethic that respects both humans and other biodiversity's rights. It should be remarked, however, that the present study showed that there are a minority (11.5%) in favour of the status quo, that is, "business as usual." They are against government interventions in the gold panning activities in the region.

Possibly they benefit (directly or otherwise) more than others from the illegal gold mining activities in the area. As such, they are afraid that once measures are put in place to control mining activities, their gains will come to an end. This is contrary to the reality that our poor, battered, plundered and polluted planet can no longer endure a continuation of 'business as usual' (see also chapter 2, this volume). It should be understood, therefore, that the earth and its wilderness though too vast can be damaged by voluntary human choice especially those that do not consider the environment as an entity that warrant respect and moral consideration. In our own time we need to revalidate the symbiotic relationship between humans, non-human entities and the environment and nurture the view that natural resources and the environment on which our lives depend should be exploited with care and in a manner that guarantees sustainability.

Concerning the government and humanitarian organizations' role on environmental management and natural resource exploitation, majority (88.4%) of the respondents indicated that there was need for provision of health facilities and education services to empower people with knowledge and sustainable ways of exploiting the gold resource in the area while maintaining the environment and themselves health. This finding concurs with results from a recent study by Africa News Network[127] in the northern provinces of Mozambique which revealed that poor farming (in Majune) could wipe away the scenic beauty of the area even before it can realize its potential as a successful

---

[127] For further discussion on environmental degradation in northern Mozambique see Africa News Network, *27/06/ 2008*. 'Mozambique green revolution will depend on small scale farmers', Mozambique Afrol News, *http://www.afrol.com/Mozambique See also* BBCNEWS: *http://news.bbc.co.uk/go/pr/fr/-/2/hi/africa/7601114.stm*

ecotourism product, as evidence of indiscriminate burning of the bush are common in the area and nobody (including the government) seems to care. In Majune, as in Manica, due to lack of government initiatives, many people by default or otherwise, are reluctant to actively take part in good environmental management during mining, hence making the state a co-conspirator in environmental mismanagement and poor resource extraction.

Having discussed study results, [negative] impacts of illegal gold panning in Manica are examined. Recommendations are given and conclusions drawn thereafter.

Impacts of gold panning in Manica

♦ *Pollution and environmental degradation*

Recent studies have shown that globally about 1. 5 million children under the age of five die each year from water and sanitation-related diseases. Following this observation, the United Nations declared clean water as a fundamental right[128] that any human being should not be denied. This entails that the right of human beings to healthy and in particular the right to clean water (for drinking and other domestic uses) has become a cause of concern for the global community. I add that the right to clean water should also be extended to nonhuman animals, plants and other such entities that in one way or another need water for their survival. Surprisingly, this concern is still a dream not only to the people in Manica province, particularly those along the rivers affected by sediment deposits from gold panning, but even beyond. In Manica province, the most affected areas include Munhena,

---

[128] See BBC News, 28/07/2010. UN declares clean water as a 'fundamental right'. Available online at: *http://www.bbc.co.uk/news/*

Mutowe, Dombe, and Mashiri. In these areas, red gold-bearing earth is being washed in the rivers, causing them to silt up, and polluting the water to the extent that safe drinking water is already becoming scarce in the region. The most affected rivers include: 'Pungue, Lucite, Revue, Zambuzi, Nhacuarara and Chimedza,'[129] Musapa, and Mufuya. These rivers which cross the districts of Manica and Sussundega have now taken up the colour of the earth in the region (reddish-brown) due to silt and clay sediments deposited by illegal gold panners. These sediments together with mercury deposited in the water during amalgamation make water in the aforementioned rivers less suitable for drinking, washing and even for agriculture. It also reduces photosynthesis rates and respiration by aquatic plants and fish respectively resulting in the death of aquatic flora and fauna.

This kind of pollution also impacts the economy of the country at large. According to a recent UNICEF Report,[130] 'water and air pollution is estimated to cost Mozambique up to 4% of Gross Domestic Product (GDP) due to the effect on the environment, health and economy'. The cost rose by 1% from the 3% of 2008, perhaps due to increased anthropogenic and commercially driven pollution in some parts of the country like Manica. The mining of gold in the province, thus, has become 'a threat to aquatic wild life,'[131] terrestrial biodiversity, human healthy and agriculture in general. Environmental degradation also remains a cause of concern. The barren, pockmarked landscapes that have developed in Musanditevera Mountains and the eastern parts

---

[129] Diario de Mozambique, 19/05/2011. Maputo: Mozambique.
[130] See UNICEF Report, 13/05/2011. "Impact of environmental degradation and emergencies on children in Mozambique-Part 2", Available @ *http://www.unicef.org/mozambique/index.html.*
[131] Opcite.

of Chimanimani and Siyawatonga Mountains have changed the aesthetic of the natural environment in these areas. I observed that gold panners in these areas sometimes dig shafts that are 10 to 20 meters (33 to 66 feet) deep and are connected by an elaborate tunnel system. What makes the shafts even more dangerous is that once exploited of their minerals they are simply left open. Besides creating mosquito breeding grounds, this leads to recurring mudslides during the rainy season causing even more serious pollution of the nearby rivers.

♦ Anti-social behaviour

In many countries, various additional problems have been associated with gold mining including child labour, gender inequity, severe health and social concerns such as HIV/AIDS and prostitution[132]. In sub- Saharan Africa, such problems have been reported in countries like Mozambique, South Africa and Zimbabwe where similar activities also take place.

In Mozambique, common health concerns include diarrhoea, injuries, sexually transmitted diseases (STDs) and parasitic infections. Mozambique had a national HIV and Aids prevalence for adult (aged 15–49) estimated at 16% in 2005,[133] but in the Manica region, the HIV rate has been reported to be one of the highest in the country (19% in

---

[132] See Report to the Global mercury project by Veiga, M.M. and Baker, R. 2004. Protocols for environmental and health assessment of mercury released by artisanal and small scale miners, GEF/UNDP/UNIDO, pp. 170.

[133] See Arndt, C. 2006. HIV/AIDS, human capital, and economic growth prospects for Mozambique, *Journal of Policy Modelling* 28 (5) 477–489.

2005)[134]. Such alarming rates have been accentuated by the fact that most of the gold fields where gold panning is taking place have become self-contained worlds where the law of the jungle prevails. And, as most of the gold miners (mainly men) leave their families behind and lead a nomadic life, they fall prey to prostitutes. Prostitutes from within, others from the neighbouring Zimbabwe, Malawi and Zambia are reported to be brought by trucks every Thursday, because mining is prohibited on Fridays[135]. Also, malaria and cholera as with HIV/AIDS are not diagnozed or treated effectively within the community as basic health necessities are absent, and most of the people involved lack education.

Besides, the government has ignored this lawlessness for as long as possible, resulting in "escalating" social violence prevailing in all illegal gold mining sites in the province. At Mutowe and Munhena, for example, violence escalates almost every evening in the camps of the gold panners. At the latter site, two of the camps have become so notorious for the regularity of their bloody quarrels that they have acquired the sinister nicknames, "Burundi" and "Bagdad" respectively. In Chazuka communal area, destruction of homesteads has also been reported[136] in cases where gold fields have been discovered at one's homestead. To make matters worse, the

---

[134] See Health Alliance International Report, 2005. Integrating TB and HIV care in Mozambique: lessons from an HIV clinic in Beira, Beira: Mozambique; Manica district of Mozambique. 2005 Final Report. United Nations Industrial Development Organization. Vienna, Austria. Available at: <www.globalmercury.org>.

[135]See Spielgel International, 24/03/2008. The gold slaves of Mozambique, Available at: *http://www.spielgel.de/international/topic/africa/*.

[136] See an article by Vimeo, 2011. "Kushata kwezvimwe kunaka kwezvimwe (Chazuka)". Available @ *http://vimeo.com/18987154/*.

governor of Manica province, Raimundo Diomba, has been quoted as resignedly saying 'after all, these people have no other source of income, and they have to feed their families'[137]. Such a statement from the respected governor has made many gold panners to think that gold panning is a legal and risk free activity.

♦ Loss of government revenues

From an economic perspective, revenue is 'income that a company receives from its normal business activities, usually from its normal business activities such as the sale of goods and services to customers'[138]. Revenues are received by the company from interest, dividends or royalties paid to it by other companies or customers. In the case of a country, revenue can be obtained from taxes paid by companies or individuals doing business in the country.

In the case of gold panning in Manica, now that a lively business has developed around gold fields in the province, the country has to collect revenues from the miners. Gold panners are reported to sell whatever they find to dealers in the country for the equivalent of about $20 (€13) per gram. The buyers in Manica are Lebanese, Israelis and Europeans, taciturn individuals who spend their evenings hanging about the city's dimly lit hotel bars, where they do their business[139]. The state's failure to make gold panners pay revenue, say in form of taxes, is therefore a total loss to government and indeed the entire country.

---

[137] Opcite.

[138] See Williams, J. R. et al., 2008. *Financial and managerial accounting*, McGraw-Hill: Irwin.

[139] Opcite.

Negotiating the problem: Some recommendations

"Gold is a natural resource and the government of Mozambique should not stop people from exploiting the resource", said one of the respondents. While the first part of this statement is true, the second part raises controversies of epic proportions especially from environment protagonists, economists, and life ethicists. In view of the observations and effects noted during the field survey carried out at mining sites in Manica province, this study recommends that the government in conjunction with the Ministries of Mines, Environment, Lands, health, and Agriculture should enter in dialogue with the illegal gold panners and establish a monitoring committee consisting of scientists and experts from different ministries and other relevant bodies to help arresting this mounting problem in the province. In particular, these authorities should cooperate in the following:

1) In assisting any potential association or company that approaches them to legally carry out mining activities in the province. This would encourage people to legally exploit the resource, hence making it easier for the government to collect revenue and control activities taking place at the mining sites in the province.

2) In constructing some settling dams for affected rivers before their confluence with the unaffected rivers. This will prevent contaminated water from passing into other [unaffected] rivers as this would extent the problem to distant areas.

3) In examining aquatic flora and fauna for mercury content before consumption by people. This is important given that mercury is a substance with the potential to bio accumulates in the food chain.

4) In facilitating human health and environmental health education or education programs in all mining sites across the province. This should be done to promote constructive symbiotic relationship between gold miners and the environment as well as to raise awareness on how the miners could reduce the spreading of different diseases in their respective areas.

To conclude this chapter, it should be underlined that the risks faced by Mozambique, and in particular Manica due to gold panning are not new and unique to Mozambique, but are resonant with other countries in the region like Zimbabwe and South Africa. However, the high levels of poverty, illiteracy and therefore the capacity of the country to deal with the effects of illegal mining activities makes Mozambique more vulnerable. Lack of infrastructure such as roads and shortage of qualified personnel like environmentalists, mining technocrats and health workers to teach and empower gold miners in Manica exacerbates the situation leading to prolonged problems in health (spread of diseases), food insecurity and environmental degradation among other challenges.

More importantly, I have argued in light of the effects of illegal gold mining in Manica that although the government is taking strides to contain the problem, its pace is too slow and in fact insignificant for the damage being caused by the activity. In view of this observation, I have proposed the need by the government of Mozambique to harmonize theory and action – need to expedite implementation of policies, improving the information circulation and movement from emergency actions to preventive plans without delay. To help this materialize, I have recommended that the government in conjunction with the Ministries of Mines, health, Environment, Lands and Agriculture should enter in dialogue with the illegal gold panners and establish a monitoring committee consisting of scientists and experts from different ministries and other relevant stakeholders to help arresting this mounting problem. This is in line with the argument elaborated in this chapter that the gold panners should be made to take into consideration all (humans, nonhuman animals, and the physical environment) those that are negatively affected by their activity.

Chapter 5

Community-Based Natural Resource Management (CBNRM) in Mozambique: Lessons and Directions for Natural Resource Management

In many countries across the world, natural resources play a significant role as sources of livelihoods for rural households,[140] and socio-economic resource base for sustenance of national economies. To unravel and understand the role of natural resources in households especially in the rural areas, we need to understand the diversity of natural resources and the nature of rural livelihoods concerned. In the case of the latter, for example, Cavendish[141] observed that it is not uncommon for a rural household to be involved in live-stock raising, growing a diversity of crops, collecting forest products for subsistence needs and sales, being involved in a variety of reciprocal transactions with fellow

---

[140] For further discussion on livelihoods see Shackleton, C.M. And Shackleton, S.E., 2000. *Direct Use Values Of Secondary Resources Harvested From Communal Savannahs In The Bushbuckridge Lowveld, South Africa. Journal Of Trop. Forest Products 6, 28-47. See also* Shackleton, C.M., Shackleton, S.E. and Cousins, B. 2001. The role of land-based strategies in rural livelihoods: The contribution of arable production, animal husbandry and natural resource harvesting in communal areas in South Africa, Development Southern Africa, 18 (5): pp. 581-604.

[141] See Cavendish, W. 2001. Rural livelihoods and non-timber forest products'. In De Jong, W and Campbell, B. (Eds). *The role of non-timber forest products in socio-economic development*, CABI Publishing: Wallingford.

community members, and involved in some small-scale industry. In terms of diversity, natural resources are diverse including minerals, fisheries, wildlife and forests. These resources make an important but complex and diverse part of rural livelihoods in many countries, especially the developing world. This is mainly because most resources in the rural areas such as wildlife, fisheries and forest products do not require high technical skills to exploit or bring them into production. This confirms Cavendish's[142] observation that there is a strong match between the characteristics of the rural poor and the characteristics of forest products and other such resources they exploit for living. In the case of forest resources, for example, there are timber and non-timber forest products, which are both important in different ways. In the case of non-timber forest products, it has been reported that wild forest products contribute as much as 35 per cent of average household income, increasing to 40 per cent for poorer households[143] in the developing world. The use of forest products are many with the most common being medicine, fuel wood, food, fodder, pillows, household baskets, sleeping mats, sponges and brooms, constructing poles, fencing materials, thatch grass,[144] chairs, walking sticks,

---

[142] Ibid.

[143] See Cavendish, W. 1999, in Shackleton, C.M. And Shackleton, S.E., 2000. *Direct use values of secondary resources harvested from communal savannahs in the bushbuckridge lowveld, South Africa. Journal Of Trop. Forest Products 6, 28-47.*

[144] See Cunningham, A.B., 1993. *Ethics, ethno botanical research, and biodiversity: People And Plants Initiative*, WWF International, Gland; See also Peters, C. M. 1994. *Sustainable harvest of non-timber plant resources in tropical moist forest: An ecological primer,* Biodiversity Support Program-WWF, Washington, D.C.; See also Chamberlain, J., Bush, R. And Hammett, A. L., 1998. *Non-timber forest products: The other forest products, Forest Product*

fish rods, and base beds. As such, FAO[145] notes that at least 150 non-timber forest products are significant in terms of international trade with the trends of trade being from the developing countries, with about 60% imported by countries of the European Union, Japan and the United States of America.

The value of non-timber forests in developing countries such as Mozambique is premised on the fact that the country has fragile agricultural and industrial systems that can hardly sustain the national economy and poor majority's livelihoods forms. In the pretext of such conditions, there is no doubt that conserving and managing natural resources in a sustainable manner will benefit Mozambique in a number of ways which include poverty alleviation and improving the nation's economic status. This is confirmed by Shackleton *et al*[146] who note that policies and programs which enhance productivity, output and incomes from natural resources have the potential to attack poverty and inequality while simultaneously promoting economic growth. On the same

---

Journal 48(10), 10-19; See also Dovie, B.D, Shackleton, C.M., And Witkowski, E.T.F., 2001. Involving local people: Reviewing participatory approaches for inventorying the resource base, harvesting and utilization of non-wood forest products, In *Harvesting of non-wood forest products: Proceedings Of FAO/ECE/ILO International Seminar*, Ministry Of Forestry, Turkey Pp. 175-187.

[145] See FAO's 1997 report on the state of the world's forests, food and agricultural organisation of the United Nations, Rome.

[146] See Shackleton, C.M. and Shackleton, S.E., 2000. *Direct Use Values Of Secondary Resources Harvested From Communal Savannahs In The Bushbuckridge Lowveld, South Africa. Journal Of Trop. Forest Products 6, 28-47.*

note, Altieri[147] emphasizes the importance of forests in general when he asserts that the forests are the sources of green spirit where traditional rituals are gathered. Human[148] elaborates Altieri's assertion by giving the example of the Kesharpur of India, the villagers whom he said have made a shrine to the goddess *'Durga'*, where villagers often perform a *puja* – an act of religious devotion, thus important on spiritual dimension to the environmentalism of the core activists in the forest. At an important conference, the United Nations Conference on Environment and Development (UNCED) held in 1992 in Rio de Janeiro, emphasized the importance of forestry with the latter receiving major attention under Agenda 21; Chapter 11 entitled "Combating Deforestation". With respect to forest use, UNCED declared the need to promote efficient utilization and assessment to recover the full valuation of goods and services provided by forest lands and woodlands[149]. All these is clear testimony that natural resources play a fundamental role in human lives. Yet, the resources can be easily stressed, become scarce and sometimes extinct if they are over-exploited or poorly managed. It is from this understanding that many countries call for sustainable use of natural resources so that the resources continue benefitting the present and future generations. The concept of sustainability re-emerged in the field of environment and conservation sciences in the late

---

[147] Altieri, M. A. 1998. *Agro-ecology. The science of sustainable agriculture,* Macmillan. New York.

[148] Human, J. 2000. *Community Forest Management: A Case From India.* Oxfam Publishers: London.

[149] UNCED. 1992. United Nations Conference on Environment and Development, Final Advanced Version of Agenda 21, Chapter 11, Combating Deforestation. Washington DC, United Nations publishers.

1980s as key to conservation and management of natural resources. Sustainability has been generally understood as meeting the needs of the present without compromising the ability of future generations to meet their needs[150].

Due to the realization that state controlled natural resource management that do not involve local communities' full participation is costly and has in many countries failed to sustainably manage resources, there has been a shift in the last two decades or so by many countries towards the promotion of community-based natural resource management (CBNRM) as externally initiated projects. Thus in conservation and management of natural resources, there has been a drift away from state-centred control to decentralization of natural resources management to allow participation of local communities in resource conservation. While the adoption and nationalization of CBNRM has become a trend in many countries in Africa and in particular sub-Saharan Africa, Mozambique is one country that is slowly moving towards that direction in spite of the efforts it made after independence to revise some of the inherited natural resource colonial legislations such as '*Legislação sobre as actividades da caça*' (Hunting activities legislation)[151]. This study examines why Mozambique has taken so long to devolve the rights to resource conservation and management to grassroots levels. The study goes beyond to explore the impact of the delay to devolve rights to resource conservation to local communities, and suggest recommendations to contain the situation. In other words, since there is a renewed

---

[150]See World Commission on Environment and Development, 1987. *Our common future,* Oxford University Press: Oxford.

[151] See Soto, B. 2003. Protected areas management in Mozambique, Report for IUCN-SASUSG, Harare.

debate on the role of different institutions in community-based natural resource management, the study further assess whether such a delay has benefited Mozambique or not, and in which ways.

Study objectives

In light of the contribution of natural resources' popularity as a source of livelihood for mostly rural communities and as a resource base for many countries' economies, natural resources present an opportunity to unravel and unlock the problem of use and knowledge in conservation of the natural resources. I, therefore, seek to unpack community-based natural resource management (CBNRM) in Mozambique as both the government and the people pursue their livelihoods and economic endeavours.

From the foregoing, it can be emphasized that this research strives to achieve the following objectives:

1) To examine the current status of community-based natural resource management (CBNRM) in Mozambique

2) To assess the contribution of CBNRM as a source of livelihood that can enhance socio-economic status of the rural people and the nation's economic base.

3) To suggest possible strategies for improving CBNRM in Mozambique.

Research questions

Primary questions

In light of the above study objectives and given that the main goal of this study is to examine the current status of CBNRM in Mozambique, the following main research question will be raised and grappled with: "What is the current status of CBNRM in Mozambique, and what can be learnt from CBNRM in Mozambique?" In order to find answers to this important question, the study will seek data that explores the following secondary questions:

Secondary questions

1. What is the perceived contribution of natural resources to socio-economic lives of the people in Mozambique?

2. What is the place of natural resources in local thought regarding the conservation of natural resources?

3. What are the key factors that lead to some communal areas residents cooperating to collectively manage natural resources in their areas, and others not?

4. What are the key lessons to be learnt from community based natural resource management in Mozambique for common property resource management in other rural areas in the region?

Framework for considering CBNRM

Sustainable development and livelihoods approach

Studying the complex nuances and subtleties between humans, biodiversity, natural resources and the state is critical to establishing sustainable social relations and network based interactions between the aforementioned actors. Unpacking the networks and relations between humans, biodiversity in the natural environment, natural resources and the state would be useful in rethinking the divisions made between knowledge and belief, the natural and the social as well as strategies or frameworks for sustainable environmental management.

This study is located within the broad framework of sustainable development in which sustainable livelihoods and community based natural resource management are part. The field of sustainable development can be conceptually broken into three constituent parts: environmental sustainability, economic sustainability and social-political sustainability[152]. Sustainable development does not focus solely on environmental or conservation issues. More broadly, sustainable development policies encompass three general policy areas: economic, environmental and social[153]. In support of this policy, several United Nations texts especially the 2005 World Summit Outcome document, refer to the "interdependent and mutually reinforcing pillars" of sustainable development as economic, social development and environmental protection. The interdependence suggests balancing the human needs against the potential that the environment has for meeting them. In view of this understanding, the term sustainable development has been defined as 'development that meets the needs and aspirations of the current generations without compromising the ability to meet those of future generations'[154]. In a more general way, the concept of sustainable development may be seen as the facilitator for balancing the conservation of nature's resources with the needs for development by human beings. That is to say sustainable development means improving the quality of human life while living within the carrying capacity of supporting ecosystems including all natural resources that human beings depend on for their livelihoods sustenance.

As highlighted above, sustainable livelihood is a constituent element of sustainable development. It is important, however, to note that sustainable livelihood drives

---

[152] See Nigerian Institute of Social and Economic Research (NISER). 2009. *Poverty Alleviation in Nigeria*, NISER, Ibadan.
[153] Ibid.
[154] Ibid.

or motivates sustainable development. For Chambers and Conway,[155] [sustainable] livelihoods comprise of the capabilities, assets (both material and social resources) and activities required for a means of living. Sustainable livelihoods relates to a wide set of issues which encompass much of the broader debate about the relationships between poverty and environment,[156] While Chambers, Conway and Scoones are right in emphasizing the existence of resources and relationships in sustainable livelihoods, their definitions falter in failing to point out that in sustainable livelihoods the relations between humans and natural resource should be one that simultaneously improves the quality of human life and the carrying capacity of supporting ecosystems. As such, I consider livelihood as sustainable when it can cope with challenges and maintain or enhance its capabilities and resources both in the present and in the future. Thus resources, both natural and otherwise, are at the centre of sustainable livelihood. No wonder why Herbinck and Bourdillon[157] note that in an attempt to make a living, people use a variety of resources such as social networks, capital knowledge and markets to produce food and marketable commodities and to raise their incomes. For Herbinch and Bourdillon, when such resources are not available or when they are undermined people tend to go under stress and shock. This is true of natural resources which when they are

---

[155] Chambers, R. and Conway, G. 1992. Sustainable rural livelihoods: Practical concepts for the 21st century, *Discussion Paper 296*, Institute of Development Studies.

[156] Scoones, I., 1998. 'Sustainable rural livelihoods, a framework for analysis', Institute of Development Studies, Working Paper 72.

[157] See Herbinck and Bourdillon, M. 2001. See also Bourdillon, M. 1987. *The Shona peoples: An ethnography of the contemporary Shona with special reference to religion*, 3rd ed, Mambo Press: Zimbabwe.

overexploited, for example, can become scarce and sometimes extinct as in the case of non-renewable resources. It is one reason why natural resources should be used in a sustainable way, hence the importance of this study. In the case of natural resource management in Mozambique, this framework helps us to unpack and understand whether the current status of CBNRM in the country sustain development in a way that is sustainably acceptable, that is, to determine if it contributes to the betterment of the livelihoods of the present people in Mozambique or not but without compromising the future generations to use the resources, and to evaluate the problems which Mozambique is facing in managing its natural resources.

CBNRM crisis in Mozambique: Lessons and directions for the natural resource management

Community – based natural resource management (CBNRM) is a concept that is difficult to pin down with precision, especially with reference to sub-Saharan Africa. This is because there are many regional typologies of CBNRM in the southern African context[158]. For this reason, there is need to clarify what we mean by CBNM in this study. This study identifies with one of the broadest definitions of CBNRM given by Murombedzi. For Murombedzi:

CBNRM defines a wide range of interventions that are designed to improve the management of natural resources in communal tenure regimes, through the devolution of certain rights to these resources, and for

---

[158] Burrow, E and Murphree, M. 1998. *Community conservation from concept to practice: A practical framework.* Institute for Development Policy and Management, University of Manchester: UK.

112

the ostensible benefit of the owners and thus managers of these resources[159].

A more or less the same definition of CBNRM has been given by Hviding and Jul-Larsen (1995) in their book, *Community resource management in tropical fisheries*, as initiatives by the state or development agents to accomplish resource management objectives through encouraging and facilitating the participation of rural communities. The above definitions of CBNRM though have the implication that the owners and managers of natural resources might be the local communities, they show that conservation and management of the resources are externally initiated either by development agents or the state and not the local communities themselves. We are left wondering as to whether local communities themselves have never initiated conservation and sustainable management of resources in the rural areas they live. For Katerere (1999) and Marongwe (2004), and rightly so, in many of Africa's rural areas has always been organic CBNRM, that is, environment conservation and resource management in the rural communities as initiated and practised by the community members themselves.

Also, as can be seen in these definitions, the key element of CBNRM is devolution of certain rights to resources by the central government to the communities at local levels. Yet devolution of rights to resources by many national governments has proven to be a difficult exercise perhaps because some politicians feel they are losing their power to

---

[159] See Murombedzi, J.C. 2003. 'Pre-colonial and colonial conservation practices in Southern Africa and their legacy today', In Whande, W.; Kepe, T. and Murphree, M.W. 2003. *Local communities, enquiry and conservation in Southern Africa*, Africa Resource Trust: Harare, pp.1.

the grassroots communities. Murphree[160] is one of the first scholars to draw attention to the problems in CBNRM when he observed that the problem with CBNRM is that many governments, although are aware of the dangers of centralization, are nervous to share authority with those at grassroots (local community members). As Murphree[161] notes, optimal conditions for CBNRM require strong tenural rights (the rural people managing resources in their communities themselves) which in turn require fundamental devolution of power, one which politicians are unlikely to make unless there is a strong political reason to do so. Thus because middle level state institutions are egoistic and power hungry, many CBNRM in southern African region have proven futile with only a few cases being successful stories. Murphree[162] thus noted that it is true that there are some areas of success, particularly in those sites in which CBNRM is organized around high market value game like elephants, lions and leopards- areas which because they have experienced high income turnover have made attitudinal changes to wildlife and conservation. Murphree was however quick to point out that such cases are few and far between such that natural resources remain threatened, and the need to conserve them more urgent than ever.

It has, however, been observed in the past decade or so that state-controlled conservation is costly and ineffective as

[160] Murphree, M.W. 1991. Communities as institutions for resource management, *CASS Occasional Paper*, University of Zimbabwe: Harare.

[161] Murphree, M.W. 1995. 'Optimal principles and pragmatic strategies: creating an enabling politico-legal environment for CBNRM', In Rihoy, E. (Ed), *The commons without the tragedy*, SADC Natural Resource Management Conference Report, SADC, Lilongwe.

[162] Opcite.

compared to CBNRM, especially in those areas where 'real' devolution of rights to resources has been accorded to local communities. As such, many countries in southern African region have resorted to the nationalization of CBNRM in an attempt to bring local communities to full participation in the conservation and management of natural resources in their areas. In Malawi and Zimbabwe, for example, CBNRM has been legitimized and nationalized in the names of Participatory Fisheries Management Program (PFMP) and Communal Areas Management Program for Indigenous Resources (CAMPFIRE) respectively. In Zambia and Namibia, CBNRM has been legitimized and nationalized in the names of Administrative Management Design for Game Management Areas (AMANDE) and Namibia Wildlife Trust with community game guards and later Namibia Wildlife Management, Utilization and Tourism in Communal Areas. South Africa does not have an acronym but its community conservation initiative, takes the form of partnerships with communities adjacent to either protected areas or private game ranches or nature reserves[163]. All these programs originally aimed at affording space for traditional leadership and the local community members to fully participate in issues of natural resource conservation and management. This has been different in Mozambique.

While since 1995, the Government of Mozambique has adopted Community Based Natural Resource Management as a strategy for the implementation of the Forestry and Wildlife policies in particular,[164] nothing much beyond that has been

---

[163] Cook, J. and Fig, D. 1995. From colonial to community-based conservation: Environment justice and the national parks of South Africa, *Society in Transition*, 31 (1): 22-36.

[164] Wirbelaeur, C., Mosimane, A.W., Mabumnda, R., Makota, C., Khumalo, A., and Nanchengwa, M. 2005. A Preliminary Assessment of

done on the ground to ensure its success and to have the program extended to other natural resources such as fisheries and mineral resources. In fact, although Mozambique has indicated that local community members should participate in the conservation and management of resources in their communities, CBNRM in the country has not been "fully" nationalized besides a few successful community management projects which take the form of CBNRM such as Ancuabe (Cabo – Delgado), Chipange Chetu (Niassa), and Sanhote (Nampula), and Pindanyanga (Manica). In fact there are little to no bold steps taken towards the nationalization of CBNRM in Mozambique. This is aptly reflected in Salomão's observations that:

CBNRM initiative in Mozambique was designed as purely desktop work solely based on literature review, and most CBNRM programs have not been subjected to any documented evaluations. The documents available on some CBNRM projects are not detailed and comprehensive enough to provide useful information such as: the number of households existing in a particular community; the social structure; livelihood activities; types of forestry resources and products and values that are associated with their extraction, processing and marketing; and how these resources impinge on the economic wellbeing of the communities. Neither was there baseline data on the basis of which the contribution of various community resource management initiatives could be evaluated. Although the Ministry of Agriculture has a sector specifically created to deal with CBNRM issues no comprehensive and systematic reports on the impact

---

the Natural Resource Management Capacity of Community Based Organizations in Southern Africa Cases from Botswana, Mozambique, Namibia, Zambia and Zimbabwe, *The Regional CBNRM Project,* WWF Southern Africa Regional Programme Office, Harare, Zimbabwe.

and current situation of the existing CBNRM initiatives throughout the country as yet exist[165].

Such inefficiency and lack of commitment and/or will on the part of the government (through its Ministry of Agriculture) to deal with issues of CBNRM can help to explain environment degradation, overexploitation of natural resources, massive deforestation and general resource mismanagement in many parts of Mozambique. It also explains why the state fails to truly devolve rights to resources to the grassroots communities.

It is against this background that one may conclude that there is CBNRM crisis in Mozambique or rather what Anstey[166] describes as a dwindling interest in CBNRM in favour of top-down management strategies such as ecotourism, trans-boundary parks, and protected areas. As Anstey remarks:

Potentially the most significant contribution of CBNRM in Mozambique is yet to come. This is particularly likely if the current emphasis on Trans-frontier Conservation Areas (TFCAs), protected areas and state-private sector dominated partnerships should generate

---

[165]See Salomão, A. 2006. 'Towards people-centred woodland management in Mozambique: can this make a difference?: community- based natural resources management in miombo forest in Mozambique and the fight against poverty', Draft Paper, Maputo, pp. 4.

[166] See Anstey, S.G. 2004. 'Governance, natural resources and complex adaptive systems: A CBNRM study of communities and resources in northern Mozambique', In Dzingirai, V. and Breen, C. (Eds). 2004. *Confronting the crisis in community conservation – Case studies from Southern Africa,* Centre for Environment, Agriculture and Development, University of KwaZulu-Natal.

demand from the community level for real changes in governance and policy, whether a result of infringement of property rights, failures in the 'parental' forms of decision making structures or a lack of tangible economic benefits[167].

Even though Anstey's assertion above is somehow comforting as it shows hope in CBNRM in Mozambique, some scholars such as Nhantumbo *et al*[168] view the crisis in resource conservation in Mozambique as something that needs immediate action to ensure its success. They also offer a generic set of recommendations on what needs to be done to make CBNRM projects in the country more effective including:

1). There is need for support for the establishment of management committees,

2). There is need for the development of guidelines for negotiation and agreement
models between the private sector and communities,

3). Mechanisms for periodical monitoring of agreements need to be in place,

4). Guidelines for accountability of the committees managing financial resources on behalf of communities need to be in place.

Yet while one can agree with Nhantumbo et al (2004) that the above recommendations necessarily have to be considered to ensure success of the CBNRM, it can still be

[167] Ibid, pp. 188.
[168] Nhantumbo, I., Foloma, M., Puna, N. 2004. *Comunidades e Maneio dos Recursos Naturais: Memórias da III Conferência Nacional Sobre o Maneio Comunitário dos Recursos Naturais*, Maputo, 21 – 23 de Julho de 2004, UICN, vol. I.

argued that Nhantumbo et al failed to include the underlying/principal reason for the failure of CBNRM in Mozambique. Identifying the source of the problem before attempting to solve it is always important. In this study, the principal reason for the failure of CBNRM in Mozambique is identified as lack of commitment and/or will on the part of the government. In fact, it's a pity to notice that nearly a decade now after Anstey's assertion above, nothing much on the part of the government has changed with regard to CBNRM. The government remains reluctant to devolve 'full' rights to resource management or at least nervous to share power and authority over natural resources with the rural people. This reluctance by the government to fully commit itself to CBNRM, and also give teeth to Land Policy and Land Law of 1995 and 1997 respectively which formalize customary rights shows that there is a conservation crisis in Mozambique. This crisis is further unveiled by the fact that the state still remains the owner of resources such as wildlife, and land and there are neither local-level institutions nor provision for this in law.

In the light of conservation crisis examined in this chapter and of course many pages of this book, there is no doubt that natural resource management in Mozambique should take a new direction. Learning from instances of success stories of CBNRM in Mozambique such as Chipange Chetu in the north, and others in the region such as Norumedzo grove (*jiri*)[169] in Zimbabwe, if carefully implemented at national level, CBNRM in Mozambique can bring back sanity to resource conservation in the country. In

---

[169] Mawere, M. 2012. Buried and forgotten but not dead': Reflections on 'ubuntu' in environmental conservation in south-eastern Zimbabwe, *Afro-Asian Journal of Social Sciences*, Vol. 3, No. 3.2 Quarter II 2012.pp. 1-20.

the case of Norumedzo *jiri*, for example, the local community members own a very big forest known as Norumedzo *jiri* and directly benefit from the resources from the forest such as edible stinkbugs (*harurwa*). In fact, the local people are "prosumers", that is, both managers and consumers of the resources they have in their community. This has allowed the resources in the area to thrive for centuries now. I therefore argue that for the CBNRM to be successful, there is need for full commitment on the part of the government. Such commitment can be demonstrated by supporting local communities with relevant and appropriate resources (both monetary and knowledge-wise) to ensure that conservation in rural communities is sustainably done. In the event that the country has no monetary resources to kick this off, it can team with civic or non-governmental organizations. A successful team-up with such organizations has been recently done in Zimbabwe where the civic organization, Southern Alliances For Indigenous Resources (SAFIRE) has helped locals in *Masawu* (merlot) fruit abundant Dande area to create linkages with Speciality Foods of Africa (SFA), provided training in business planning and management of the *masawu* fruit that have helped to meet some of the challenges the group faced[170]. The civic organization, SAFIRE, which focuses on community development using non-timber forest products, *masawu*, works with the local community in the harvesting, processing and commercialization of *masawu* jam. SFA buys the fruit directly from communities in Zambezi valley with the assistance of SAFIRE and Speciality Foods for Africa and SAFIRE's cooperation ensures that a fair price is paid to the communities and that the highest quality fruit is

---

[170] See SAFIRE, 2003. Annual Report 2002, Harare. SAFIRE. Available @ http://www.safireweb.org).

used for the jam[171]. With such interventions and assistance from civic organizations, the Dande community members are motivated to conserve and sustainably manage their merlot forests.

The other important thing that the government of Mozambique should do is to devolve rights to natural resources to the local government and local communities as has been explained in the Norumedzo case given above. Devolving rights to natural resources by the government would ensure full participation, accountability and responsibility of local community members in the conservation and management of resources in their respective areas. As observed by Bond,[172] if people are managerially integrated and enjoy the material benefits of conservation, they become motivated not only to conserve natural resources but to put in place institutional mechanisms for this purpose. Besides, devolving rights to local levels also gives local communities a sense of ownership of the natural resources they manage and at the same time enjoy the benefits obtained from the resources. As argued by Nkhata,[173] real devolution is important because it provides people not

---

[171] Ibid.

[172] See Bond, I. 1999. 'Economic incentives for institutional change for the management of natural resources', In Johnson, S. And Mbizvo, C. (Eds). *Proceedings of the exchange visit workshop for directors*, IUCN-ROSA, Harare.

[173] See Nkhata, A. 2004. 'Devolution and natural resources management in Zambia: Transforming rural communities into gatekeepers without authority', In Dzingirai, V. and Breen, C. (Eds). 2004. *Confronting the crisis in community conservation – Case studies from southern Africa,* Centre for Environment, Agriculture and Development, University of KwaZulu-Natal.

only with management authority but also with ownership of resources as where management authority is devolved; it transforms communities from gamekeepers (decision implementers) to real managers (decision makers). As Nkhata remarks:

CBNRM is failing because of the limited devolution of critical aspects of natural resource management from the centre to the periphery. The decision making and benefit distribution process from government to rural areas has not been adequate enough to allow CBNRM deliver the [intended] goods[174].

Anstey[175] gives a good example of a project in northern Mozambique, Chipange Chetu where at its inception the local community members were unwilling to take part mainly because they felt the project was 'foreign' and so would not directly benefit them. Thus Anstey observes that one of the major obstacles to the institutionalization of community based forestry initiatives in Chipange Chetu was the fact that many of the local people lacked confidence in the project. Anstey thus notes:

Even after eight months of discussion locally and the implementation of a number of activities such as the PRA process, discussion of objectives and controls over outsider use, there was still a general belief that the program was for the benefit of the Government or individuals in the

[174] Ibid, pp. 64.

[175] Anstey, S.G. 2001. 'Necessarily vague': The political economy of community conservation in Mozambique', In Hulme, D. and Murphree, M.W. (Eds). *Community conservation in southern and eastern Africa,* James Currey: Oxford.

NGOs and the talk of devolution to local institutions was merely a variation on a historical theme of local disempowerment[176].

The third thing that the government can do to ensure that CBNRM in Mozambique is successful is to strike a balance between all stakeholders involved in the conservation and management of natural resources in the country. This means a reversal of the current scenario in Mozambique where the state and private business seem to benefit most from resources such as timber, wildlife, minerals and fishery while the local community members who live near and are in fact owners of these resources languish in abject poverty. As such, there should be a balance in terms of shares between all stakeholders involved in conservation and management of natural resources, be they private businesses, the state or local community members. Where there is no such balance and where some stakeholders are excluded, they get frustrated and do whatever is possible to downplay efforts of those in the forefront of conservation and management of natural resources. This explains why in most cases the government or non-governmental interventions in environment and resource conservation fail as long as they exclude the local community members from the front seat in terms of decision making and implementation of whatever projects they design around conservation.

In short, this chapter has examined community-based natural resource management in southern African region but with particular reference to Mozambique. It has established that though indications are there that Mozambique is somehow willing to fully implement CBNRM – where either organic CBNRM or CBNRM as externally initiated, in the

---

[176] Ibid.

conservation and management of its natural resources – there seem to be lack of commitment and/or will on the part of the government to legitimize the strategy at national level. A clear sign or indication for this reluctance has been that CBNRM in Mozambique, unlike in other countries in the southern African region, has not been 'fully' nationalized. As such, there has been gross mismanagement of natural resources in many parts of the country as is demonstrated in many chapters of this book, a situation I may consider as a state of conservation crisis in the country.

In the face of this crisis in the country, a number of recommendations have been offered for the country to adopt to ensure effective and successful management of natural resources. It has been suggested, for instance, that the government of Mozambique should provide local communities with the appropriate managing resources such as knowledge, capital, and most importantly devolve rights to natural resources to the local government and local communities in that order. Devolving rights to natural resources by the government would ensure full participation, accountability and responsibility of local community members in the conservation and management of resources in their respective areas.

Chapter 6

'It's Neither Conservation Nor Preservation' - Possibilities for Alleviating Poverty Through Dambo Utilization in Chokwe Rural area, Gaza Province

Mozambique, like many other countries in Africa experiences a number of problems ranging from poverty, corruption, poor governance and low productivity. While these problems are interconnected that we cannot address one of them without addressing the others, because of space and for purposes of this study, I focus more on the problem of poverty including extreme poverty. Poverty in Mozambique, especially in the rural areas has stood out as a key indicator of the cataclysmic socio-economic challenges that have characterized Mozambique for decades now. The 16 years of civil unrest, political tensions and entanglements after national independence in 1975 in Mozambique exacerbated an already precarious socio-economic and political landscape in the country. This contributed to extreme state fragility and ushered in a cocktail of economic crisis in Mozambique. Mozambique thus, can be characterized as a fragile state. OECD defines fragile states as 'those where the state power is unable and/or unwilling to deliver core functions to the majority of its people: security, protection of property rights, basic public services and essential infrastructure'[177] leading to

---

[177] See OECD cited In Makochekanwa, A. and Kwaramba, M. 2009. 'State Fragility: Zimbabwe's horrific journey in the new millennium', *A Research Paper Presented at the European Report on Development's (ERD)*, Accra: Ghana, pp. 4.

socio-economic tumultuous situations even more than two decades after civil war in Mozambique.

Besides the blame laid on the political episode that followed immediate after national independence from the Portuguese colonial government, the economic crisis in Mozambique has also resulted from high levels of corruption, over-exploitation and sometimes underutilization of resources. All these factors have had catastrophic consequences on the national economy, the long struggling agricultural sector (since the colonial period), and the lives of the already poorest and most disempowered masses in the rural areas. What is both worrying and surprising is that the overwhelming majority of Mozambique's population affected by poverty resides in rural areas where land, a basic means of production/survival and source of income is not only abundant but rich in other resources. This is echoed by *Censo-*census[178] and Rambe and Mawere[179] who noted that: '71% of Mozambique's total population lives in the rural areas' with the geographic distribution of the population being that many of these people live in remote, rural areas where land and other resources are abundant. A critical question arises here: "If majority of Mozambique's population lives in the countryside where natural resources and land are abundant, why the country remains poor?" While this question will be responded to later in this chapter, it is worth noting that poverty in Mozambique, as elsewhere in the developing world, is predominantly a rural phenomenon, though

---

[178] *Censo* (Census), 2007. Maputo, Mozambique.

[179] See Rambe, P. and Mawere, M. 2011. Barriers and constraints to epistemological access to online learning in Mozambique Schools, *International Journal of Politics and Good Governance*, 2 (2.3 Quarter III): 1-26.

according to reported data Africa has experienced the highest level of rural-urban migration relative to other continents, in part as a result of policies which have discriminated against agriculture and promoted industry[180]. Poverty constrains the general development capacity of the rural people as they do not have access to the essentials to improve their lives on a sustainable basis; poverty deeply threatens the country's development at all levels. Also, poverty sometimes cultivates "mental laziness" by inculcating a dependency syndrome in the affected people's minds. Such factors compromise the government and significant success of the country's socio-economic development efforts as mental laziness is likely to result in a higher form of poverty I shall call mental poverty. Mental poverty is the poverty of the mind gotten when due to dependency syndrome one becomes too lazy to think of possible alternatives to better his [her] own and other people's lives. Possibly, because of the dependency syndrome nurtured by donor agents since the liberation struggle through civil war, among other factors, Mozambique remains one of the poorest nations in the world with more than 80% of its population in rural areas live on less than a US$1 a day, and lack basic services like schools and hospitals[181]. Yet while it is generally agreed that the majority of Mozambicans live in poverty, the concept of poverty especially extreme poverty is difficult to define with precision as it keeps on changing. In

---

[180] Lele, U. and Adu-Nyako, K. 1991. 'An integrated approach of strategies for poverty alleviation: a paramount priority for Africa', A paper prepared for the annual meeting of the African Development Bank Group, Abidjan: Cote d'Ivoire.

[181] For further discussion on poverty in Mozambique see Rural Poverty Report-Mozambique, 2007. Maputo, Mozambique; See also Afrol.com, 2008. Available online @ *http://www.afrol.com/Mozambique*.

this study, I identify with the World Bank's 2005 and 2010 definition that 'extreme poverty is living on less than US$1.25 a day. This meant living on the equivalent of US$1.25 a day in the US buying US goods. In 2010, this meant surviving on the equivalent to US$1.50, AUD$2 or 1 pound per day'[182]. Thus, according to this definition, majority of Mozambique's population live in extreme poverty as they live on less than US$1 a day, and lack basic services like schools, roads, clean water and hospitals.

That said, this chapter critically reflects on the utilization of natural resources, and in particular how dambos can possibly be exploited to alleviate poverty in the rural areas in Mozambique. The chapter advances the argument that the extreme poverty that have haunted Mozambique, especially since post-colonial independence is partly a result of poor and under-utilization of natural resources such as dambos, among other reasons. Granted, the continued existence of poverty is partly a result of prevailing socio-political and economic situation in Mozambique which arguably is responsible for cultivating mental poverty (and also material poverty) among the rural people, and for creating opportunities for a plethora of vices such as corruption. This is predicated on the adage that poor governance breeds misdeeds. From the foregoing, it can be argued that good governance and sustainable exploitation of natural resources such as dambos that have been laid idle for decades now can help alleviate or eradicate extreme poverty in Mozambique. This argument is premised on the fact that extreme poverty in Mozambique is predominantly a rural phenomenon where natural resources such as dambos are abundant yet people lack innovative

---

[182] For more on extreme poverty see World Bank. 2010. Extreme poverty rates continue to fall, Available online @http://www.worldbank.org.

thinking to explore the resources for their common good. As such, an effective way to alleviate poverty in Mozambique seems to depend on the country's ability to increase opportunities for income generation and sustainable utilization of natural resources such as dambos by the rural majority where this resource is widely available. This way, dambos in Mozambique can possibly be used as "bridging zones" for the integration and alignment of locally generated knowledge and scientific knowledge through creating liberal spaces for the rural people's experimentation with different forms of practices (in dambos) that foster sustainable development and aims at eradicating extreme poverty.

Understanding the concepts of 'dambo' and 'poverty'

The concept of dambo though gained prominence in intellectual discourse over the years in Africa and beyond, has not been easy to define with precision. A number of interpretations have been conjured by different scholars. Some scholars, have for instance defined 'dambos' – also termed mbugas, vleis and fadamas – as seasonally saturated, grassy, channelless, gently sloping valley floors that commonly occupy the lowest topographic positions in African catenae or 'land systems' [183]. In more or less the same

---

[183] For further discussion on dambos see Watson, J.P. 1964. A soil catena on granite in southern Rhodesia, I. Field observations. *Journal of Soil Science*, 15: 238-250; See also Acres, B. D. et al. 1985. African dambos: their distribution, characteristics and use. In Thomas, M.F. & Goudie, A.S. (eds.), *Dambos: small channelless valleys in the tropics.* Zeitscrift für Geomorphologie, Supplement band, 52: 63-86; See also Ollier, C.D. et al. 1969. *Terrain classification and data storage: land systems of Uganda,* M.E.X.E.

way, Bullock[184] defines dambo as one of the several dialectal terms used to describe the seasonally saturated, grassy, generally treeless narrow depressions covering as much as 20% of the plateau regions of central and southern Africa. Though the two definitions elaborated above seem to be somehow different, what underlies the two is that dambos are part of a range of habitat types known as wetlands. In some quarters, particularly in sub-Saharan Africa, dambo also known as "*bani*" (in Shona of Zimbabwe) and "vlei" (in Afrikans-South Africa) has been defined as a Bantu term used to describe the extensive seasonally saturated, grassy depressions common to central and southern Africa[185]. It is worth mentioning that while most dambos are waterlogged for the larger part of the year, majority of them dry out especially at the surface during winter season (where they receive rainfall in summer). As noted by Mackel,[186] the "sponge-like" centre of most dambos stays moist even during the dry season, sustaining the higher-density herbaceous vegetation in this zone. This connotes that the distinctive characteristic feature of dambos is that they conserve moisture even during dry season, but while they are waterlogged for the larger part of the year, most of them dry

---

Report No. 959, *Military Engineering Experimental Establishment,* Christchurch, Hampshire, U.K.

[184] Bullock, A. 1992. Dambo Hydrology in Southern Africa – Review and Reassessment, *Journal of Hydrology*, 134: 373-396.

[185] For definition of dambo across different cultural groups you can see http://www.geo.utah.edu/dambo/index.html.

[186] Mackel, R. 1985. Dambos and related landforms in Africa – an example for the ecological approach to tropical geomorphology. In M.F. Thomas & A. S. Goudie (Eds.), *Dambos: small channelless valleys in the tropics.* Zeitscrift für Geomorphologie, Supplement band, 52: 1-23.

out especially at the surface during winter season (where rain falls in summer). For this reason, dambos are also characterized by dense grassy vegetation.

As noted by another group of scholars, the other important characteristic feature of dambos is that they have a relatively planar topography ($\frac{1}{2}$ – 2° slope) which makes them produce little hydraulic energy, facilitate soil saturation and inhibits channel formation[187]. This agrees with the observation that dambos are relatively flat (1/2 – 2° slope), which inhibits drainage and the formation of streambeds[188].

Like the discourse on dambo, the discourse on poverty is highly momentous and has sustained controversies of epic proportions in Mozambique's discussions on development studies. Given the tenuous nature of the concept of poverty coupled with the different interpretations evoked by the deployment of the concept across different contexts and situations, a robust comprehension of the concept that calls into question its practical manifestations and application in particular situated contexts is imperative. In Mozambique, the tenuous nature of the concept of poverty continues to haunt researchers in spite of the fact that since the country's independence from Portugal in 1975, one of the major concerns of the national government and donor agencies has been to fight poverty. It seems, therefore, that the emergency of the concern to fight hunger has been largely a result of

---

[187] Acres, B. D. et al. 1985. African dambos: their distribution, characteristics and use. In Thomas, M.F. & Goudie, A.S. (eds.), *Dambos: small channelless valleys in the tropics*. Zeitscrift für Geomorphologie, Supplement band, 52: 63-86; See also Von der Heyden, C. J. 2004. The hydrology and hydrogeology of dambos: a review. *Progress in Physical Geography*, 28: 544-564.

[188] See http://www.geo.utah.edu/dambo/index.html.

observed considerable effects of certain economic reform programs in the country such as Economic and Structural Adjustment Program (ESAP), and not a deeper and nuanced understanding of what poverty entails. Without closing possible definitions by future researchers, I will make some attempts here to understand the concept of poverty before spelling out its effects and how it can be possibly eradicated in some rural areas of Mozambique.

Though the phenomenon of poverty is not something unique and new to Mozambique and even the world's history, it should be underscored that a quick glance at the relevant literature shows that there is no general consensus on any meaningful definition of poverty[189]. Ogwumike, for instance, defines poverty as a household's inability to provide sufficient income to satisfy its needs for food, shelter, education, clothing and transportation[190]. It should be acknowledged that Ogwumike's definition captures the important indicators of poverty such as the inability of a household to satisfy its basic needs. However, his definition falls short for the reason that it fails to include the aspect of healthy which is also a critical aspect in the definition of poverty. Such a limitation is also notable in the former President of the World Bank, Robert McNamara's definition. McNamara,[191] defines poverty

[189] For further discussion on the conceptualization of poverty see Schubert, Renate. 1994. Poverty in Developing Countries: its Definition, Extent, and Implications. *ECONOMICS* FRG, Vol. 49/50; See also Nigerian Institute of Social and Economic Research (NISER). 2009. *Poverty Alleviation in Nigeria*, NISER, Ibadan.

[190] Ogwumike, F. O. 1991. A Basic Needs Oriented Approach to the Measurement of Poverty in Nigeria, *NJESS*, Vol. 33, no 2, 105-119.

[191] McNamara, 1995. In World Bank, "Ghana poverty past, present and future", *Report No.14504-GH*, Washington DC: USA.

as a condition of life so degrading as to insult human dignity. While McNamara's understanding captures the notion of poverty, it doesn't shed more light on how poverty as a condition of life degrades and insult human dignity. A more precise definition and aspects of poverty is perhaps captured in The Ninth Report of the Development Policy of the Federal German Government which states that people affected by poverty are unable to lead a decent life. The report elaborates on how people affected by poverty are unable to lead a decent life by listing the following critical aspects of poverty:

Poverty means not having enough to eat, a high rate of infant mortality, a low life expectancy, low educational opportunities, poor drinking water, inadequate health care, unfit housing and a lack of active participation in decision – making processes[192].

In this book, the above definition by the Development Policy of the Federal German Government shall be adopted for the reason that it captures most if not all critical aspects of poverty, some of which the aforementioned definitions by McNamara and Ogwumike left out by default or otherwise.

---

[192] See BMZ, 1992. Federal Ministry of Economic Co-operation and Development: Ninth Report on German Government Development Policy, Bonn, pp. 13.

Problem background and geographical description of the study area

Mozambique is a country that is found along the western coast of the Indian Ocean. As such, its greater part especially in the southern provinces is covered by dambos. This study was not conducted in the whole of Mozambique, but in southern Mozambique and in particular Chokwe district of Gaza province using Chokwe rural area as a case study. Chokwe is located close to the Indian Ocean, on the southern side of Limpopo River, which run through the province emptying into the Indian Ocean near Xai-Xai city. In terms of agro-ecological regions, Chokwe is in region 2. It is about 230 km north of the capital city, Maputo, and is in a wide, fertile plain where rice and tomatoes are grown though not at a large scale. Chokwe which shares boarders with Bilene and Xai-Xai districts in the south and east respectively, has a population estimate of 61 666[193]. The relief of the area is generally flat and is characterized by vleis and isolated trees. The altitude of the area is about 30 m above sea level but with some low lying areas below sea level. The area has warm to hot summers and cool winters. The mean annual temperature is about 30°C. The rainfall though falls throughout the year, depending on the year and influence from the sea, it is mainly conventional and occurs between November and March. The climate of the area is the tropical type as it includes one rainy season, one dry season; it is hotter during dry season and cooler during wet season. The mean annual rainfall for the area is about 500 – 750 mm per year with vegetation consisting of a Tropical dry savannah.

---

[193] The population figures for Chokwe district, Mozambique is available at: http://en.wikipedia.org/wiki/chokwe.

It is worth noting that though the larger part of the communal area is covered with dambos, only a smaller part of the dambos was converted into an irrigation scheme (for wheat and tomato commercial farming) during the Portuguese rule. The Portuguese left the scheme immediate after independence in 1975 giving it back to the black Mozambicans from who the land had been taken away during colonial era. Unfortunately, the scheme fell in the hands of the *Estado Novo* (new state). Thus as Tanner[194] noted, the history of land occupation in Chokwe is complex and contentious. The history illustrates a long-term process in which local people have experienced one round of land dispossession and repossession at the hands of the Portuguese, and a second round of dispossession and repossession at the hands of the FRELIMO government. Yet as a researcher on environment and development related issues on southern Africa, I have come to the realization that most researchers on dambos devoted their attention to the general description of dambos[195] and land disputes,[196] without

---

[194] Tanner, C. 1993. 'Land disputes and ecological degradation in an irrigation scheme: A case study of state farm divestiture in Chokwe, Mozambique', *Paper presented to Land Tenure Centre*: University of Wisconsin-Madison.

[195] For general description of dambos see Acres, B. D. et al. 1985. African dambos: their distribution, characteristics and use. In Thomas, M.F. & Goudie, A.S. (eds.), *Dambos: small channelless valleys in the tropics*. Zeitscrift für Geomorphologie, Supplement band, 52: 63-86; See also Von der Heyden, C. J. 2004. The hydrology and hydrogeology of dambos: a review, *Progress in Physical Geography*, 28: 544-564; See also Mackel, R. 1985. Dambos and related landforms in Africa – an example for the ecological approach to tropical geomorphology. In M.F. Thomas & A. S. Goudie

tackling some specific important issues, for example, exploring how dambos can be used as a drive towards poverty alleviation. The research on dambos in Africa and in particular Mozambique thus makes a sorry reading with its failure to document, by default or otherwise, the various ways through which dambos can be exploited to alleviate poverty among the rural populations. The consequence is that dambos remain an idle resource either poorly utilized and/or underutilized by the rural majority. The dambos potential to eradicate or at least alleviate poverty of the rural population remains underestimated and therefore undocumented. This is the situation in which people in the Chokwe rural are currently in.

One more important thing to note is that during the pre-colonial era, the Chokwe people practiced shifting cultivation, hunting and gathering in these dambos. Traditional vegetables and legumes like *coleus esculentus*, corchorus olitorius, amaranthus gracilis, ipomea aquatic forsk, momordica balsamina and sonchus oleraceus (*tsenza, guche, m'boa, terere, chidledlelane, cacana/ncaca and chinhamucaca, nlhavi* respectively in local Shangani),[197] among others, were mainly grown for domestic consumption. Further, it was revealed by my informants that in the pre-colonial era, wild animals like dambo mice, *mavhondo,* duiker, jaguar, hares and some birds were hunted, and fruits like *hute* (*syzigium cordatum*) coconut and cashew nuts harvested. This scenario changed

(Eds.), *Dambos: small channelless valleys in the tropics.* Zeitscrift für Geomorphologie, Supplement band, 52: 1-23.
[196] Opcite.
[197] Tembe, M. J. 2008. 'Indigenous vegetables and legumes' importance, utilization and marketing in Gaza Province, Mozambique', Available online at:
http://www.mct.gov.mz/pls/portal/url/ITEM/5017142A159E9D4CE040007F01004B9.

dramatically with the Portuguese's settlement and *prazor* (commercial farms) system in the area of course the whole of Mozambique. The area that occupies Chokwe was designated to commercial activities, mainly rice growing. This saw the Chokwe people losing their land to the Portuguese settlers, only to get it back after independence.

Having looked at problem background and geographical location of study area, the next section of this study focuses on the method(s) employed to carry out the research.

Methodological issues

This study was carried out in 2010 in Chokwe rural area in southern Mozambique. As part of my research design, I relied on field observations, literature studies, informal discussions and interviews. A selected sample of 36 people (18 female and 18 male) were conducted during the research. The sample size of 36 was considered sufficient in providing the general perceptions of the Chokwe people on the utilization of dambos in their area. I observed some of the activities that take place in the Chokwe dambos such as harvesting of traditional vegetables, cattle pasteurization, fishing (in some remnant pools) and hunting. The field observation was used to ascertain the actual activities that take place in the studied area. To supplement data obtained through field observation, interviews were conducted with a view to get more information on how the dambos were being utilized, and to determine if the dambos were fully utilized or underutilized by the local people and why.

Participants for this study were drawn from different classes of the society that comprises Chokwe rural area, ranging from the highly educated to the least educated and the working to the non-working class. Drawing participants from across Chokwe rural area was done with the hope to obtain a balanced research result that speaks for the whole area. The age group of the participants ranged from 15 to 65 years. This age group was considered appropriate for the study for the reason that most of the people involved in direct use of natural resources and other issues related to economic development in the Chokwe area are within the aforesaid age range. Equal number of men and women were sampled for the mere reason that both sexes should be equally represented and participate in socio-economic issues that affect their community. I administered questionnaires with both open and closed items (open questionnaire and closed questionnaire) to the participants in the different areas they were found. The open questionnaire was used as it enables the respondent to reply as he [she] likes and does not confine the latter to a single alternative[198]. In fact open questionnaire evokes a fuller and richer response as it possibly probes deeper than closed questionnaire by moving beyond statistical data into hidden motivations that lie behind attitudes, interests, preferences and decisions (see chapter 2, this volume). On the other hand, the closed form of questionnaire was used because it facilitates answering and makes it easier for the researcher to code and classify responses especially in this case where a large number of questionnaires were to be dealt with. Both questionnaires (open and closed) were used because in practice, a good questionnaire should contain both open and closed forms of questions so that responses from the two forms can be checked and compared[199]. The participants responded to

---

[198] Behr, A. L. 1988. (2nd Ed). *Empirical Research Methods for the Human Sciences*, Durban Butterworths.
[199] Ibid.

questionnaire items individually and participation was voluntary. Participants were also assured of the confidentiality of their responses and to show the authenticity of the assurance, participants were asked not to identify themselves by names. Collected data were tabulated to show frequencies before being subjected to evaluative analysis. The Tables 5 and 6 respectively contain details of the people participated in the study and the data that was gathered during the study:

Table 5: Participant demographics

Occupation	Gender	
	Male	Female
Cattle headsman	3	2
Village head	2	1
Fisherman	2	1
Peasant Farmers	5	7
Students in public education	2	2
Hunters	2	1
Students in tertiary education	2	2

Table 6: Responses to closed questionnaire items

ITEM	RESPONSES		
	Agree	Disagree	Uncertain
1. There is extreme poverty in Chokwe communal area.	32	4	0
2. Dambos are useful resources to human livelihood	35	0	1
3. Dambos in Chokwe are underutilized.	25	10	1
4. Dambos in Chokwe communal area should remain idle and unexploited.	4	35	1
5. Dambos can be used to alleviate poverty in Chokwe communal area and beyond.	20	10	6
6. Dambos can be used for a wide range of economic activities.	25	10	1
7. Dambos should be used only for hunting and grazing.	4	31	1
8. Dambo utilization should be controlled to ensure sustainable use and development.	28	6	2
9. The number of Agritex officers in Chokwe communal area (if any) should be increased.	30	5	1
10. There should be government and non-governmental initiatives on dambo utilization.	30	5	1

Discussion based on research findings

The results in Table 6 show different perceptions on the utilization of dambos by the Chokwe people. This will be discussed later in this section. I will quickly note that majority (88. 9%) of the respondents confirmed that there is extreme poverty in Chokwe communal area. This concurs with the Rural Poverty Report-Mozambique[200] which noted that Mozambique is one of the poorest nations in the world and more than 80% of its citizens in rural areas live on less than a US$1 per day, and lack basic services like schools, hospitals, clean water and roads. Yet poverty in Mozambique is a tale of two cities as there are both extremely poor people and very rich people in the same society. This economic inequality has been made possible by the fact that wealth in Mozambique is unfairly distributed. It [wealth] is in the hands of the minority elite group who sometimes benefit and continue to enrich themselves through unscrupulous or corrupt means. This is because in Mozambique, corruption is one of the most prevalent problems affecting both government institutions and local communities. The phenomenon of corruption is aptly captured by Mackenzie who in her studies of forestry and logging industry in Mozambique had this to say:

The timber [in Zambezia province of Mozambique], part of it sold cheaply and illegally through corrupt means by some local business people or government officials, is exported as unprocessed logs, undermining local industry and transferring most of its potential benefits from one of the poorest countries in the world, to what is becoming one of the richest. Together with local business interests and Asian traders, these public

---

[200] Rural Poverty Report-Mozambique, 2007. Maputo, Mozambique; See also Afrol.com, 2008. Available online @ *http://www.afrol.com/Mozambique.*

servants constitute a "timber mafia". Instead of combating illegal logging, they are, through measures including the manipulation of forest regulations, technical information and statistics, accepting bribes and personal involvement in logging, actually facilitating and personally benefiting from this "Chinese takeaway"[201].

Perhaps such acts as those described in the extract above together with lack of information on poverty levels in the country explain why 11. 1% of the respondents strongly disagreed that there is no extreme poverty in Mozambique. The unfair distribution of wealth in Mozambique is confirmed by Hanlon who noted that 'besides assistance from international organizations and the 4. 5% economic growth of Mozambique, the number of poor people in the country is ever increasing'[202].

Dambos and the Chokwe people's livelihood

On whether dambos are an important resource to the Chokwe people's livelihood, an overwhelming majority (97. 2%) agreed. This is echoed by Tembe[203] who notes that Gaza province (in which Chokwe is located) is home to indigenous vegetables and leguminous species such as coleus esculentus, corchorus olitorius, and amaranthus gracilis, among other

---

[201]Mackenzie, C. 2006. Forest governance in Zambezia, Mozambique, Mozambique: Chinese Takeaway! *Final Report for FONGZA,* Maputo, pp. iv.

[202] See Verdade Jornal-Newspaper, 04/12/2009. Maputo, Mozambique.

[203] Tembe, M. J. 2008. 'Indigenous vegetables and legumes' importance, utilization and marketing in Gaza Province, Mozambique', Available online at: http://www.mct.gov.mz/pls/portal/url/ITEM/5017142A159E9D4CE0 40007F01004B9

things that grow naturally in the vleis. This was further confirmed by one of the respondents who remarked that "it is a historical truism that dambos are an important resource to the people in this area as most of us always depend, in one way or another, on the dambos for survival". As such, majority (86. 1%) agreed that dambos should not only be used for grazing and hunting, but for some other activities that have higher socio-economic values.

Dambo conservation or preservation?- Panacea for poverty alleviation

It should be underlined that there were positive and negative perceptions with regard to dambo utilization in the Chokwe. This means that there were mixed feelings among respondents in Chokwe on how dambos were and should be used. Such mixed feelings were very much visible on the question whether dambos should be either preserved (remain untampered with) or conserved, that is, exploited in a sustainable manner that would alleviate poverty in Chokwe. I observed that 55. 6% agreed; 27. 8% disagreed, and 16. 7% were uncertain. Reasons for these mixed perceptions on this important question were varied but the major one was that people in the Chokwe area have different levels of education, ambitions, and economic status. Students in tertiary education, for example, had a strong conviction that dambos can be conserved, that is, protected while at the same time used to sustain the lives of the people in the district. It, however, came out clearly that some locals confused conservation with preservation. One of the respondents, Tania, a female student at Universidade Pedagogica, Gaza aptly clarified the difference in his comment:

I strongly believe that the people's lives around here can improve greatly if the people are furnished with expertise on sustainable utilization of dambos. The problem is that some of us think that preserving, that is, leaving the dambos lying idle for future generations without exploiting them at all is better than conserving them- protecting while at the same time exploiting the resources for sustenance. Others are also just lazy to think about what they can do with these dambos to conserve the resources while making a living out of them.

From the findings of this research, I concluded that those who disagreed and were uncertain either lacked knowledge on how dambos can be conserved and sustainably used to sustain their lives, too reluctant to make a living out the dambos, or they were too poor to exploit the dambos. One of the participants, for example, remarked: "We have always been poor with these dambos around us. I, therefore, disagree that dambos can alleviate our poverty in any way."

There were also mixed responses on whether dambos in the Chokwe area were being underutilized. Majority (69. 4%) agreed. Among these was one village head known as *Chefe de Quartrao* who remarked: "It is undeniable that our dambos here are being underutilized as one can easily see that the larger part of them is just idle. I believe we have never utilized this resource fully since we settled here". As a researcher, I was agreeable to this remark as the first thing I observed when I first entered this area was the vast piece of idle dambos stretching from the south to the north and from the east to the west. The dambos seemed not to contain neither aesthetic nor economic value to the locals besides serving as sources of their livelihood. Perhaps this explains why 27. 8% of the respondents disagreed and 2. 8% were uncertain. This result was agreeable with the result on whether dambos can be used for a wide range of economic activities were 69. 4% agreed but still 27. 8% disagreed and 2, 8 % uncertain.

It was clear from my findings that local communities in Mozambique, as elsewhere, are not without voice or agency in the struggle against poverty as while a larger part of the dambos was idle, majority (97. 2%) agreed that dambos should not remain idle when in fact people are languishing in abject poverty. Action was required, yet extreme poverty and lack of knowledge on how best dambos could be utilized to improve the lives of the locals were revealed to be drawbacks. There was need, therefore, for some form of initiative especially from the government to make 'things' happen. It is from this realization that the majority (83. 3%) of the respondents agreed that there was need for government and donor agencies/non-governmental initiatives on dambo utilization. Agricultural Extension Officers (Agritex Officers) who can educate locals the good farming methods were also cited by the majority (83. 3%) as a requirement even if the government and non-governmental organizations were to come in with financial assistance. Those who were uncertain (2. 8%) and disagreeable (13. 9%) were perhaps hopelessly affected by their poverty. They were no longer optimistic that any kind of intervention (either from the government or otherwise) can alleviate their poverty and change their lives for the better. They had resigned to their fate. This was observed in one of the respondents who bitterly remarked: "We are now tired of empty promises from politicians and no longer have faith that our situation here can be ever improved in any way. We were born and bred in poverty and so can't hope for anything better". This sign of hopelessness is a clear testimony that extreme poverty affects both the physical and psychological states of individuals to the extent that it cultivates yet another form of poverty-mental poverty perpetuated by rhetoric of the government and nurtured by annual donations from non-governmental organizations.

The results discussed above clearly show that a lot more is desired to be done in Chokwe to boost locals' morale and improve their lives, socially and economically using some of the locally available resources in the area such as dambos. The study also revealed that dambos in Chokwe are neither preserved (protected without being utilized) nor sustainably conserved (protected while being used to sustain livelihoods for present generations but without compromising capacities to use the same by future generations). The next section focuses on recommendations that might be useful for the Chokwe people to consider if they are to alleviate their poverty using the resources – dambos – at their disposal.

The way forward for dambo utilization in Chokwe: Some recommendations

As previously highlighted, the discussion of results obtained from this study have indicated that serious action is required if extreme poverty in rural areas such as Chokwe is to be alleviated or eradicated using the locally available resources-dambos. It was clear that people in Chokwe require civic education/intellectual empowerment on how they can best sustainably exploit their major resource-dambos-to alleviate poverty. Yet such an initiative to uplift the entire community and mitigate poverty can only be taken by concerted effort of both the government and donor agencies. This can be referred to as the first phase in the elimination of extreme poverty. It should be noted that this first step initiative is critically important as it is now generally believed that most integrated projects fail chiefly because they are too complex and try to do too much too quickly as well as based on very little knowledge of the precise constraints the poor households face[204]. Cognizant of this realization, it is argued here that conscious participation of the local population in projects that directly affect their lives and communities is necessary if meaningful and sustainable development is to be attained.

After the first phase, financial aid to help the locals to start projects and for technological change (appropriate technological advances/purchased inputs that suit constraints faced by the local community) could now be issued out. Projects that can be done in the Chokwe dambos are many. These include: brick moulding, poultry, piggery and the growing of food crops that do well in dambos such as sugarcane, banana, rice, tomatoes (in mid dry season), wheat and yams. Such projects have been successfully launched in some parts of Mozambique where dambos are found. A good example is brick manufacturing in Magude district of Maputo province where locals are sustaining their families by commercially manufacturing earth bricks in the dambos. An interview with Pedro, one of the people involved in the Magude project, revealed that though they lack sophisticated brick moulding machines, the business is quite viable. He thus remarked, "Our project is doing very well. It's only that we do not have advanced machine to facilitate work and maximise our profits. For this reason, we are failing to meet demand as some of our customers come from as far as Maputo to purchase the bricks". Rice and tomatoes are also doing very well in the dambos around Chokwe town where

---

[204] See Lele, U. and Adu-Nyako, K. 1991. 'An integrated approach of strategies for poverty alleviation: a paramount priority for Africa', A paper prepared for the annual meeting of the African Development Bank Group, Abidjan: Cote d'Ivoire.

irrigation farming by the state, though at a small scale, is taking place. So is the sugarcane plantation in the nearby Chinavane dambos. In fact, Chinavane sugarcane plantation is practically the source of raw materials for all sugar mills in Mozambique. All these successes in other dambos in Mozambique clearly show that the possibility of alleviating poverty in the Chokwe area through sustainable dambo utilization is very high if careful initiatives are undertaken. Such projects as some of those named above, though practised at small scale are important in poverty alleviation because they act as the engine of national economic growth through the generation of socio-economic growth linkages.

Another way of helping the Chokwe people out of their extreme poverty through dambo utilization is to introduce Agricultural Extension Officers (Agritex Officers) in the area. According to the information provided by some informants during fieldwork, there is currently no Agritex Officer working in the Chokwe area. Agritex Officers would help promote participation of the rural community especially in farming activities by equipping the people with knowledge, skills and advices on different projects related to farming. This is what scholars like Abrams[205] advocated when he argued that in community-based projects the community should control the project and make important decisions, although professionals such as engineers may provide expertise and finance may be provided by external financial sources. This initiative thus would not only boost production in the dambos, but also encourage sustainable utilization of the dambos and participation of the rural population in

---

[205] See report by Abrams, L. J. 1996. 'Review of Status of Implementation Strategy for Statutory Water Committees', Department of Water Affairs and Forestry, Pretoria.

development issues; it promotes the rational use of available resources and maximize production and household incomes.

Third, there is need to improve infrastructure in Chokwe rural area and establish an efficient viable marketing environment to enhance agricultural growth as well as growth of other such projects as earth brick manufacturing. As Lele and Adu-Nyako[206] noted, rural infrastructure development needs to be accorded top priority. For the duo and indeed so, rehabilitation and maintenance of rural roads are essential for transporting the expected surplus to markets; incidents of agricultural produce rotting in remote parts of a country for lack of road infrastructure or because the roads are impassable are common in Africa. This is to say that as long as there is poor infrastructure (as I observed in the Chokwe area during field research), then even if the first and second recommendations are satisfied, it will remain difficult for rural producers to maximize their returns. Yet, improving the human capital of the rural poor will help create productive employment opportunities that outpace high population growth rates and therefore alleviate or eradicate their (the rural poor) poverty.

Fourth, it should be remarked that while the recommendations discussed above have the potential to eradicate extreme poverty and induce growth in the national economy, it is still possible for any benefits of these programs to bypass the targeted group – the poor. Generally speaking, this may be accounted for by factors such as natural disasters, wars, political instability, mass migrations, corruption and illiteracy, among others. However, in Mozambique the major factor seems to be corruption. This is because corruption in the country has become a chronic problem affecting almost

---

[206] Opcite.

all government and private sectors as well as many other social systems. Mozambique was ranked joint 130th out of 180 countries in the 2009's edition of the Corruption Perception Index (CPI), published by the anti-corruption NGO, Transparency International (TI)[207]. Such a scale of corruption is, therefore, a cause for alarm as corruption is normally a symptom of bad governance and structural weaknesses with the potential to undermine Mozambique's future socio-economic development. This means that if the suggested recommendations are to be successfully implemented and catastrophic consequences prevented, there is need for bold steps and tangible political commitment by the government to eradicate corruption. Yet to achieve this would also require the government to significantly increase its accountability and efficiency by passing new by-laws and establishing new institutions (like anti-corruption programs) aimed at stamping out corruption. Once these are in place, it becomes possible to indiscriminately prosecute trespassers who divert and facilitate leakages of subsidies, among other offences.

In summary, this chapter has argued that the poverty problem faced by Mozambique especially by people in the rural areas as a result of natural resources underutilization, among other factors, is not unique to Mozambique, but resonant of many developing countries in Africa and beyond. However, high levels of corruption, illiteracy and bad governance make Mozambique more vulnerable. Lack of infrastructures such as roads and shortage of qualified Agriculture Extension Officers to educate/empower rural communities and especially peasant farmers in the rural areas

---

[207] For more details on corruption in Mozambique see allAfrica.com. 2009. 'Mozambique: Country performs poorly on corruption index'; Available online at: http://allAfrica.com/.

even aggravates the situation leading to the perpetuation of extreme poverty in many of Mozambique's rural areas. In view of this observation and the findings obtained from the study, it has been argued that the Government of Mozambique and donor agencies should make concerted effort to educate its citizens (through civic education), stamp out corruption, and improve infrastructure in rural areas to ensure efficient transportation of both inputs and outputs in the area. In this study, Chokwe becomes a representative of all other rural areas in Mozambique and beyond facing similar problems.

More importantly, the chapter has argued that there is possibility of alleviating poverty in Mozambique's rural areas such as Chokwe through increased utilization of natural resources available, in this case the dambos. Yet for this possibility to materialize the study has recommended government and donor agencies to empower the poor rural people intellectually and financially. Such initiatives are important as they eradicate mental/intellectual poverty (which is the most dangerous form of poverty) before dealing with material or socio-economic poverty. The initiatives also promote the people at grassroots to actively participate in resolving problems that directly affect them such as poverty and corruption. The positive effect of actively involve local communities in development projects in their own areas is aptly noted by Winarto[208] who has this to say of Indonesia:

Indonesia's integrated pest management program has increased not just the participants' confidence in their own farming decisions but also

---

[208] See Winarto, cited In Dove, R. M., Sajise, E. P., and Doolittle, A.A. (Eds). 2011. *Beyond the sacred forest-Complicating conservation in southeastern Asia*, Duke University Press, London, pp. 20.

their willingness to question the efficiency of government agricultural programs and extension assistance.

Exporting the assertion above to the Chokwe context, it clearly demonstrates that integrating the local communities in proposed projects in Chokwe may help eradicate both intellectual and material poverty among the people.

Chapter 7

Small-Scale Cash Income Sources in Central Mozambique: The Root Cause of Resource Scarcity and Environmental Instability?

The larger part of central provinces of Mozambique, namely Manica and Sofala is endowed with diverse natural resources such as wildlife, forests, fisheries and minerals. Like in other rural areas in southern African region, these resources remain the major source of livelihood for the majority of the rural poor in the aforementioned provinces. Recent study by Shoko reveals that:

More recently, small-scale mining and alluvial gold panning activities have taken centre stage as a result of both the economic structural adjustment programs (ESAPs) and recurrent droughts within the SADC region. It is estimated that up to two million people directly or indirectly benefit from small-scale and alluvial panning of minerals within the Zambezi Basin[209].

In central Mozambique as in the Zambezi Basin of Zimbabwe, many rural people engage in gold panning as their source of livelihood. Others depend on traditional beer

---

[209] For further discussion on gold mining in Zambezi Basin see Shoko, S.M. 2002. Small-scale mining and alluvial gold panning within the Zambezi Basin: An ecological time-bomb and a tinderbox for future conflicts among riparian states, In Chikowore, G., et al., (Eds). *Managing common property in an age of globalization-Zimbabwean experiences,* Weaver Press: Harare, pp. 67.

brewing, fishing, small-scale logging, charcoal production and selling, firewood sales, hunting, fruit sales, grain crop sales, livestock sales and casual labour, but all at small-scale. Such small-scale cash income sources though play a fundamental role in enhancing the socio-economic status of the rural population in many countries, some of them are detrimental to the natural environment. In fact in many cases where activities such as artisanal gold mining, charcoal production, firewood selling, hunting and fishing are not controlled by the government through responsible ministries, they result in environment degradation, deforestation, siltation, pollution (water and land), reduction and sometimes extinction of some animal species and so on. Taking an example of the impact of gold panning on rivers, Shoko aptly observes that:

Most of this mining activity takes place on riverbeds and banks and releases enormous amounts of silt and heavy metals into river systems, dams and lakes. Siltation of rivers reduces river conveyance and the storage capacity of reservoirs, which in turn makes several areas prone to flooding[210].

In view of such studies and observations made elsewhere in the Southern African region, this research sought to assess the potential impact of small-scale cash income sources of the people around a rural township, Dombe in rural Mozambique. Dombe is located in Sussundenga district in the Manica province of central Mozambique. The location was found suitable for this study owing to the diverse small-scale cash income sources available for the people living around this township, which include traditional beer brewing, artisanal gold mining, fishing, charcoal production and selling,

---

[210] Ibid.

firewood sales, hunting, fruit sales, grain crop sales, livestock sales, earth brick moulding and casual labour.

From the findings of this study, it was established that most of the households rely heavily on hunting, gold panning, charcoal production and farming as important components of cash income or sources of livelihood, yet the government does virtually little to ensure that the people exploit the resources they depend on in a manner that is sustainable and less threatening to the natural environment, the present and future generations.

Theoretical framework – Sustainable livelihoods

This study is situated in the theoretical framework of sustainable livelihoods. The sustainable livelihoods framework has over the years been applied in various case studies or situations attempting to establish a link between resource access by the rural populations, impact of access to resources by the rural people and environment policy[211]. Technically, livelihoods entail the means, activities, entitlements, assets by which people do make a living through natural or biological means (i.e. Land, water, common property resources, flora, fauna), social (i.e. community, family, social methods, participation, empowerment) and human (i.e. knowledge, creation of skills) and are therefore

---

[211] For further discussion on the link between resource access and rural populations see Hobley, M. and Shields, D. 2000. 'The reality of trying to transform structures and processes: forestry in rural livelihoods'. *Working Paper 132*, Overseas Development Institute, London; See also Nicol, A. 2000. Adopting a Sustainable Livelihoods Approach to Water projects: Implications for Policy and Practice. *Working Paper 133*, Overseas Development Institute, London.

paramount to the debate on sustainable development. For some scholars, 'livelihoods' consist of five capitals namely human capital (knowledge, skills, health and labour), natural/physical capital (soil fertility, forest resources, water resources, grazing resources and land), financial/produced capital (credit, savings, remittances, agricultural implements and infrastructure), social capital (leadership, rules, social organization, adherence to rules, mutuality of interest, kin and ethnic networks), and cultural capital (cultural practices, traditions and identity maintenance). As such, the concept of sustainable livelihoods is an integrating concept, incorporating social, economic, cultural and ecological dimensions[212]. Consisting of resources and capacities, the sustenance of livelihoods could make a significant contribution in alleviating or even eradicating poverty whilst protecting environmental resources[213] for now and for the future generations. For Campbell and Luckert, 'a livelihood is sustainable when it can cope with, and recover from, stresses

---

[212] For further discussion on sustainable livelihoods see Carney, D. 1998. *Sustainable rural livelihoods: What contribution can we make?* Department for International Development, London; See also Bebbington, A. 1999. Capitals and capabilities: A framework for analysing peasant viability, rural livelihoods and poverty, *World Development*, 27: 2012-44; See also Cain, J., Moriarty, P., Lynam, T., and Frost, P. 1999. An update on the integrating modelling strategy, Micro-catchment management and common property resources, 2nd Integrated Modelling workshop, Institute of Environmental Sciences, University of Zimbabwe and Institute of Hydrology, UK.

[213] See World Commission on Environment and Development (WCED). 1987. *Our common future,* Oxford University Press: Oxford; See also Chambers, R. and Conway, G.R. 1992. Sustainable rural livelihoods: Practical concepts for the 21st century, *IDS Discussion Paper 296*, Institute of Development Studies, University of Sussex, Brighton.

and shocks, and can maintain or enhance its capabilities and assets both now and in the future while not undermining the resource base'[214]. In view of this understanding, livelihoods can only be considered sustainable if they can help to improve lives of the people at present without compromising those of future generations – what can be termed sustainable livelihood. This is to say that sustainable livelihoods as a framework for research takes an open view of the combinations of different resources and activities that turn out to constitute a viable livelihood strategy for the rural population. As argued by Rakodi and Loyd-Jones,[215] such an approach is crucial to the analysis of what the poor themselves do to survive in various environments as it provides a conceptual guide to think about objectives, scope and priorities of rural development as a framework for designing policies and practical interventions and their subsequent evaluation in poverty reduction. To this end, it can be argued that for livelihood to be sustainable there is need to seriously consider the following factors which have direct effects (negative or otherwise) on the resources being used by the rural people:

---

[214] Campbell, M. B. and Luckert, K. M. (Eds). 2002. *Uncovering the hidden harvest- Valuation methods for woodland and forest resources,* Earthscan Publications Limited, London, pp. 2002: 14.

[215] Rakodi, C. and Loyd-Jones, T., 2002. *Urban Livelihoods: A People-Centred Approach to Reducing Poverty.* Earthscan Publications Ltd: London.

♦ Diversification of resources

For rural livelihoods to be sustainable there is need for diversification in the resources used in the countryside by rural populations. As observed by Ellis, diversification can be loosely understood as the process by which rural families construct a varied portfolio of activities and social support capabilities in order to survive and improve their living standards[216]. As further noted by Ellis, a diversified livelihood comprises, for example, a number of activities which are on-farm or off-farm and income generating or non-income generating for survival or accumulation with the former done by the poor and the latter by the rural rich and include but not limited to intensive agriculture and or non-agriculture.

♦ Sustainable survival strategies

In rural areas where many people do not have access to productive land or jobs, poverty is prevalent and harvest or extraction of natural resources can provide means to barter for commodities needed for daily survival[217]. When the rural people harvest or exploit resources in their areas, they normally employ strategies in exploiting the resources. Scoones[218] defines livelihood strategies as comprising of agriculture intensification or extensification, livelihood diversification and migration. For Benedikz, livelihood

[216] Ellis, F., 1998. Survey article: Household strategies and rural diversification. *Journal of Development Studies*, 35(1): 1-38.

[217] Campbell, M. B. and Luckert, K. M. (Eds). 2002. *Uncovering the hidden harvest- Valuation methods for woodland and forest resources,* Earthscan Publications Limited, London.

[218] Scoones, I. 1998. Sustainable rural livelihoods, a framework for analysis. Institute of Development Studies, Working Paper 72.

strategies can be viewed as coping or adapting strategies[219]. Coping strategies are activities that redress short term surges in stresses and shocks while adapting strategies are long term and tend to be more resilient[220]. This means that livelihood strategies are ways employed by the rural populations when exploiting resources for use. Yet while under 'normal' circumstances the rural people employ those strategies or ways that ensure the continued existence of the resource, things become different when factors such as poverty and lack of government interventions come into play. Put differently, the use of strategies is determined by a number of factors including status i.e. whether the users are poor or rich, seasonality, coping behavior, and intermittent investments and savings, among others. Whatever the case, survival strategies by resource users should always be sustainable to guarantee continued existence of resources for both the present and the future generations.

♦ *Educational capacities*

By educational capacities, I mean the general level of understanding or formal education of a people that can assist them in the exploitation of resources around them. This means that educational capacities are an important component for sustaining livelihoods although in many developing countries there are still high illiteracy rate as not all the people have access to formal education. To effectively and meaningfully involve communities in livelihood strategies

---

[219] See Benedikz, K., 2002. Rural Livelihoods at Risk: Land use and coping strategies of war-affected communities in Sri Lanka, *Proceedings on the Conference on International Agricultural Research for Development*, Witzenhausen, Deutscher Tropentag, Oct., 9-11.
[220] Ibid.

that are sustainable therefore demand for the provision of knowledge and skill to the resources users. The Dombe case study presented in the following sections looks at a range of resources strategies and activities undertaken by the community, with and without proper knowledge and skills to sustainably extract as well as use the resources.

While it has been brought to light that the framework used in this study is sustainable livelihoods, the study does not attempt to critique the framework in terms of arguments for or against it. The objective is to provide a detailed impact assessment of rural livelihoods, particularly at community and household levels. The latter has been applied in Mozambique and other Southern African countries and it clearly shows the impact of policies on the management of natural resources and household economies[221]. The strength of the sustainable livelihoods framework used in this study is that it critically examines the involvement of community at the micro level, that is, at the household level in fishery, forestry, mineral and wildlife related activities in the context of a variety of other livelihood strategies that they undertake, such as gold panning, firewood selling and charcoal production, among others. The impact of resources constraints such as availability of clean water, land for cultivation, access for harvesting, and wildlife commercial purposes, and so on, is analysed against the different activities that the family undertakes. The analysis of the impacts of resource constraints provides insights for environmentalists and policy makers.

---

[221] See Nhantumbo, I., Dent, J.B. and Kowero, G.S. 2001. Goal Programming: application in the management of the Miombo Woodlands in Mozambique. *European Journal of Operational Research* 1332 (2): 310-322.

Having looked at the theoretical framework of this study, the next sections discuss the case study and methodologies used in carrying out the research.

Dombe case study: Competing resource uses

Dombe is a business centre in Sussundenga district of Manica province, about 80kms from the provincial capital, Chimoio. There is wide range of resources available for the people around the business centre. Due to the fact that Dombe is a fairly big township in a rural area, there is stiff competition for resources among the rural people around the business centre. The people exploit the resources and sell them at a usually ready market at the township. Other activities like artisanal gold mining, large numbers of road workers regarding the road that passes through Dombe, forestry and tourism, further exacerbate competition for resources in this area thereby accrediting the political ecologists'[222] criticism against the assertion that rural communities are solely to blame for routinized environmental degradation and poor resource extraction. With the realization that external forces at national or international or both levels normally exert enormous pressure and stress on resources in the rural communities, political ecologists have shifted the problematizing gaze of environmental degradation and [unsustainable] resource extraction from the livelihood practices of local peoples to the consumption patterns and

---

[222]Neumann, P. R. 1998. *Imposing wilderness: Struggles over livelihood and nature preservation in Africa*, University of California Press: Berkeley; See also Doolittle, A. 2007. Native and tenure, conservation, and development in a pseudo democracy: Natural resource conflicts in Sabah, Malaysia, *Journal of Peasant Studies*, 34 (3): 474-97.

economic growth of the wider national and even global community, exposing the wider political, economic and social forces that influence local decisions regarding resource use[223]. In the context of Mozambican rural areas such as Dombe, the political ecologists' argument holds much water as some utilization of resources in the area is externally driven. This is exemplified by the case of gold mined in the artisanal mine fields around Dombe which is purchased by some buyers and tourists from abroad. As previously highlighted the major livelihood sources available for people at and around Dombe include gold panning, farming, charcoal production, fishery, beer brewing, hunting, and livestock breeding. Yet overlapping interests and exploitation of resources can be a major source of not only conflict, but competition and unsustainable use of resources if proper planning and government intervention are not carried out and agreed upon between different stakeholders and government agencies. For example, discussions with local communities, and people affected by gold panning at Musanditevera mountains in the nearby Mutowe rural area (in Manica) indicated that they were strongly opposed to the government's plan of establishing Mining Associations and allowing uncontrolled mining activities in the area, because of their conflict with the existing land use plans and lack of sufficient measures to protect community interests.

---

[223] See Dove, R. M. 1993. The responses of Dayak and Bearded Pig to Mast-Fruiting in Kalimantan: An analysis of nature-culture analogies, In Hladik, M.C. (Ed). *Tropical forests, people, and food*, UNESCO: Paris; See also Doolittle, A. 2007. Native and tenure, conservation, and development in a pseudo democracy: Natural resource conflicts in Sabah, Malaysia, *Journal of Peasant Studies*, 34 (3): 474-97.

Methodological issues

This study was carried out in 2012 in Manica province, particularly around a business centre known as Dombe Township. Three major approaches in livelihoods research namely the retrospective approach, the circumspective approach and the prospective approach were used. As Murray[224] notes, the retrospective approach entails looking back in the past experiences of the household` life trajectories to understand long term changes in livelihoods, for example, a year or so back, while the circumspective approach focuses on what households are currently doing to survive, and prospective approaches entail prospective approaches to livelihood studies. Though the retrospective and prospective approaches remain important, circumspective approach is more central to this study as it provides what the households are doing currently to survive, cope with socio-economic constraints. As such, the study used the household as the basic unit of observation, source of data and analysis. A household in this case, is considered to be a social group which resides in the same place, shares the same meal and makes joint or coordinated decisions over resource allocation and income pooling[225]. The household head, representative (or any household member) is the respondent for the selected households. Respondents were thus randomly selected before interviewed and administered to some questionnaires to verify household data. Informal discussions and questionnaires were the basis of data collection to get critical,

---

[224] Murray, C., 2002. Livelihoods research: Transcending boundaries of time and space. *Journal of Southern African Studies*, 28(3): 489-509.

[225] Ellis, F., 1998. Survey article: Household strategies and rural diversification. *Journal of Development Studies*, 35(1): 1-38.

informative and statistical data about resource use in the area. The research used purposive random sampling in selecting households' respondents for administering both in-depth interviews and questionnaires. The results from field observations, interviews and questionnaires were compared before subjected to critical content analysis. This triangulation process helped me to verify inconsistencies in data gathered during research. Content analysis was used on qualitative data while quantitative categorical data generated were coded according to themes and tabulated as below for purposes of clear and easy presentation and analysis.

Table 7: Participant demographics

Occupation	Gender	
	Male	Female
Livestock dealer	3	3
Charcoal Producer	4	2
Fisherman	2	1
Peasant Farmer	5	5
Firewood seller	4	4
Hunter	4	1
Gold panner	4	2
Beer brewer	2	4

Discussion based on research findings: On the possible impacts of unsustainable small-scale income sources to the environment

The findings presented in this study are based on the data that were collected from participants at Dombe township and others living around the township between September and October 2012. Besides data obtained from field observations and questionnaires, informal interviews were used as data gathering techniques. As previously noted, the principal objective of the research was to obtain comprehensive information about the impacts of small-scale cash income sources in Mozambique's rural areas, particularly to the natural environment.

Table 8: Responses to closed questionnaire items

ITEM	RESPONSES		
	Agree	Disagree	Uncertain
1. Most of the local people around here depend on natural resources in this communal area as their sources of livelihood.	49	1	0
2. There is dwindling of some resources in this area.	50	0	0
3. The dwindling of some resources in this area is a result of competitions for the resources by the users, and unsustainable exploitation i.e. over-exploitation and poor extraction of such resources as minerals, forests etc.	40	8	2
4. Some small-scale income	38	10	2

sources such as charcoal production and fuel wood sales have long term negative impacts to the natural environment i.e. causing pollution, deforestation etc.			
5. Some people around here over-exploit or badly extract resources due to ignorance and poverty.	36	3	11
6. To avoid straining livelihood sources available there is need for resource diversification.	39	6	5
7. Many people who depend on small-scale income sources for their livelihood lack 'good' strategies that sustain the sources for longer periods.	36	3	11
8. There should be a control system for the utilization and extraction of resources in rural areas where most of these resources are found.	30	18	2
9. There are no Environment Monitors in Dombe communal area, and if any they should be increased.	50	0	0
10. There should be government and non-governmental initiatives on natural resources, most of which are used as income sources.	30	18	2

An additional objective was to explore possible solutions to the problems being caused by the unsustainable extraction and exploitation of natural resources by those who use them as small-scale cash income sources.

The research results in Table 8 above reveal that majority (98 %) of the respondents confirmed that natural resources are an important cash income source for most of the people in the Dombe area, as in many other rural areas in Mozambique and beyond. However, there were different perceptions on the impacts of small-scale cash income sources by people in the area. The main problem faced by resource users in Dombe was the dwindling of some resources they use as their small-scale cash income sources. The dwindling resources included firewood, charcoal and fish. This dwindling has mainly been a result of stiff competition for the resources between users. According to majority of the participants (80 %), there has been mounting resource crisis and a scramble for the resources in the recent years. Besides competition by the locals themselves, the scramble for resources in Dombe has been exacerbated by the existence of road workers working on the road that passes through the area. The road construction company, which resumed its operations in 2011, has a large number of workers mainly from outside the Dombe community. This has in turn [negatively] impacted on the available resources and caused overexploitation and skyrocketing of prices of some resources in the area. Situations such as this, though observed at global level, are also noted by scholars such as Chibisa and Rwizi[226] who observe that the current

---

[226] See Chibisa, P. and Rwizi, L. 2009. Traditional Crafts and Rural Livelihoods in Manicaland's Semi-Arid Areas: Implications for

skyrocketing of global food prices is due to poor harvest resulting from climate change, rising demand for bio fuels and market speculation. As the duo speculated, the skyrocketing of global food prices means that rural communities especially in developing countries such as those in sub-Saharan Africa will become more dependent on forest/woodland food sources.

In terms of the way resources are being extracted in the area, the research revealed that poverty, ignorance and lack of resource diversification and poor government interventions are the main factors that contribute to the unsustainable use of resources. Majority (72 %) of the respondents, for example, agreed that people over-exploit or poorly exploit resources in their area due to ignorance and poverty. Due to poverty and economic hardships, rural parents cannot raise the required money for school fees and textbooks for their children to use at school and home respectively. As a result, they resort to [natural] resource extraction by whatever means at their disposal. In fact, where resources are scarce, there is competition for those little but strained resources. In such contexts, over-exploitation and poor resource extraction practices normally persist and the resources' capacity to serve the people severely compromised. This kind of situation was confirmed by the findings of this research which revealed that majority of the respondents (78 %) registered the need for diversification of resources to afford them the opportunity to choose or look for other resources as their sources of income. In the light of the limited resources available to the Dombe people, the latter are forced to compete for those resources available to get income to sustain their families.

There was also indication that no Environment Monitor (EM) was available for the Dombe people to consult whenever they encounter any environmental problem or even to educate the 'local' people the importance of sustainable use of natural resources and how this can be done. This was echoed by majority (100 %) of the respondents who expressed concern on the unavailability of EMs in their area. As one respondent elaborated, the absence of such members as EMs or at least government control mechanisms on resource exploitation meant that people could extract and use the resources available wily nilly. The respondent's concern also helped me to explain my field observations, particularly on why water in the nearby Lucitu River, which is a perennial river and important source of livelihood for the people, had turned reddish-brown. I was told that the water had changed its colour due to artisanal gold mining activities in the nearby Musanditevera Mountains. The 'panners/illegal minners' do their mineral processing in the Lucitu river thereby polluting the water. The other indication that there were no Environment Monitors in the area was the intensity at which stream bank cultivation along Lucitu and Musapa Rivers, and deforestation (for fire wood and charcoal production) just opposite Dombe Business Centre, across Lucitu River where charcoal production is done and with harvesters targeting quality wood. This poses adverse ecological and environmental effects. As Mackenzie noted:

The focus of harvesting on a small number of species inevitably affects the overall composition of the forest, and the removal of the best quality timbers of individual species inevitably changes the genetic composition of their populations, and some species may suffer permanent reduction in their populations…..Other main commercial species are being left with insufficient number of immature trees to ensure their survival and the basic integrity of the forests. For example, species such as mondzo, whose populations are dominated by very large very old trees, or pau preto and pau ferro, which can be taken for timber at lower diameters and are also exploited by communities for pole wood, are particularly vulnerable[227].

In relation to stream bank cultivation, this has long term effects such as siltation and environmental degradation. Although some of the plants grown along the river such as sugarcane and banana are resistant to erosion, this study revealed that the natural ecosystem along the Lucitu and Musapa Rivers was being destroyed. More so, the recommended 30m distance from the river bank was neither followed nor known by most of the local farmers who engaged in stream bank cultivation (along Lucitu and Musapa). The long term effects noted above were also far from known by most of those involved in stream bank cultivation.

Challenges for addressing problems to do with resource use

From the foregoing, it is clear that the livelihood sources for the people in Dombe area though at small-scale are straining the resource base where these resources are obtained: the people are causing stress and shocks to the resources. Yet there are factors identified as contributors to the unsustainable utilization of resources in the area. These include:

---

[227] See Mackenzie, C. 2006. Forest governance in Zambezia, Mozambique, Mozambique: Chinese Takeaway! *Final Report for FONGZA,* Maputo, pp. 73.

♦ Low educational levels

Mozambique is one of the countries in the world with high illiteracy rate and poor access to other educational means such as internet that help enhancing educational capacities of a people. Since 1993 when internet was introduced in Mozambique 'only 1.6% of its total population has access to internet'[228]. Worse still, 'of the 1.6% who have access to internet, 75 % of them live in the capital city, Maputo where the population is only 1.109.798 ('08),'[229] and the people there do not directly depend on forest, non-timber forest products and other such resources for their livelihood. Rural populations that constitute the majority in the country still suffer from virtually no access to electricity and computers. Bearing this in mind, there is no doubt that rural areas such as Dombe run the risk of severe land degradation, water shortages, high pollution levels, deforestation and climatic changes if no immediate action is taken to arrest the situation. As highlighted in the discussion above, this study revealed that ignorance is one of the problems that leads to over-exploitation and poor extraction of resources in Dombe. This means that as long as people in the Dombe area are not educationally capacitated, over-exploitation and poor extraction of resources in the area will persist. In view of this problem, the government and concerned non-governmental organizations should take some steps towards educating the rural people, through environmental awareness campaigns, about the importance of sustainable utilization of resources.

---

[228] For further discussion on internet use and access in Mozambique see World Internet Statistics, 2010. Mozambique.
[229] Ibid.

171

♦ Poverty

The other factor that continues hindering sustainable utilization of resources in Dombe area as in other such rural communities in Mozambique is poverty. Mozambique remains one of the poorest nations in the world and with more than 80% of its citizens in rural areas live on less than a US$1 a day, and lack basic services like schools and hospitals[230]. As highlighted in the discussion above, poverty remains a stumbling block for sustainable utilization of resources in the rural areas as people are left with no option (for other resources) except to compete with other users for the available resources. This is not to say that sustainable livelihoods in rural areas are impossible, even if the government commit itself. The argument advanced here is that for this possibility to materialize there is need for the government and donor agencies to empower the rural people intellectually and financially. The government through its Action Plan for the Reduction of Absolute Poverty (PARPA) program should source funds to fight poverty in the country, especially in the rural areas where majority of the population live. Such initiatives are important as they fight both mental/intellectual poverty (which is the most dangerous form of poverty) and socio-economic poverty. Once such initiatives are made the local community members will begin to look for better alternative sources of income thereby actively participate in resolving environment related and socio-economic problems that directly affect them such as poverty.

---

[230] For more information about poverty levels in Mozambique see Rural Poverty Report-Mozambique, 2007. Maputo, Mozambique; See also Afrol.com, 2008. Available online @ *http://www.afrol.com/Mozambique*.

♦ Unemployment

Due to poverty and poor industrial base, Mozambique remains one of the countries with the highest unemployment rate. According to CIA,[231] Mozambique's unemployment rate is still pegged at 60 % though there are indicators that it may fall to lower levels. The findings of the present study also revealed that most of the people involved in environment threatening activities such as gold panning, stream bank cultivation, charcoal production and fuel wood selling in the Dombe area were not formally employed. As such, the activities they engaged in were their sole sources of income. Yet such activities coupled with erratic rainfalls and the global economic recession may exacerbate the environment related problems in Mozambique. To arrest the situation, I therefore urge the government to create more job opportunities at community level (using its resources such as forestry, fishery and mineral, among others) for its people so as to reduce the number of natural resource users in the rural areas. Mackenzie[232] stressed the same point when she advised that since Mozambique is a developing country, it needs to harness all its resources to achieve its social and economic objectives. Forestry, minerals, fishery and wildlife represent some of the few resources in rural areas with real economic potential, if community development is integrated closely with these resources. This can be done, for instance, by

---

[231] For further discussion on unemployment rates in Mozambique see CIA, 2009. List of Countries by Unemployment rate, Available at: https://www.cia.gov/library/publications/the-world factbook/fields/2129.html.

[232] See Mackenzie, C. 2006. Forest governance in Zambezia, Mozambique, Mozambique: Chinese Takeaway! *Final Report for FONGZA,* Maputo, pp. 73.

widening the country's industrial base such as the logging industry or negotiate with non-governmental organizations (NGOs) specializing in rural livelihoods to help people in the rural areas on issues to do with environment management and natural resources exploitation. A case in point for the latter is a civic organization in the neighbouring country, Zimbabwe, known as Southern Alliances for Indigenous Resources (SAFIRE). This organization has helped the rural people in the rural Dande to create linkages with another organization, Speciality Foods of Africa (SFA), provided training in business planning and management that have helped to meet some of the challenges the rural people in the Dande area were facing[233]. Such interventions have had positive effects on the lives of the local people and in the conservation of a vast forest of *Masawu* (merlot) fruit trees. In 2003, for example, the project raised close to ZWD 2 million from sales[234]. In view of these examples, it can be argued that similar interventions should be made in the rural areas of Mozambique such as Dombe so as to alleviate poverty and save the natural environment from further deterioration through over-exploitation and stress from competition by users. Otherwise, the environment threatening activities such as deforestation, charcoal production, fuel wood selling, and gold panning remain a time bomb that, in the future, will aggravate Mozambique's environmental as well as socio-economic problems.

In short, this chapter has examined the impacts of small-scale cash income sources in central Mozambique to the natural environment. The chapter has revealed that

---

[233]SAFIRE, 2003. Annual Report 2002, Harare: SAFIRE. Available at : http://www.safireweb.org.
[234] Ibid.

conservation issues can in no way escape politics. To this effect, it has been established that majority of the populations in rural Mozambique, and in particular the Dombe area, largely depend on activities such as gold panning, wildlife resources, farming, charcoal production and fuel wood selling for survival, resources which the nation as a whole partly depend on. Yet due to factors such as poverty, competition for resources and low education capacities, most of the resources in area are either over-exploited or poorly extracted/harvested. This creates room for environmental harms such as deforestation, exhaustion of some resources, pollution and siltation, among others. The latter problems, for example, result from stream bank cultivation being practiced along Lucitu River. In view of these observations and findings, this study has suggested the need to explore alternative development paths and resource bases for Mozambique's rural communities facing increasing pressures on their natural resources. This suggestion has been proffered on the understanding that in an area with limited resource capacity or competition for resources, an increasing demand of resources can lead to unsustainable development and other environmental ills such as land degradation, pollution etc. To avoid such ills and to maintain sustainability in the natural environment, the need to look for alternative development paths and introduction of other sources of livelihoods that help reduce shocks and stresses on available resources become necessary.

More importantly, the chapter has recommended that the government of Mozambique should launch environmental awareness campaigns and introduce Environment Monitors in the rural areas. Such initiatives have been argued to have the capacity to change attitudinal behaviour of the rural people. Still on that, the government has been urged to create employment for the rural people so as to reduce competition and the number of people who rely on the natural resources in the countryside.

Chapter 8

Unlocking the Crisis: New Directions for Conservation in Mozambique

Conservation has generated and continues to generate vigorous debate about how the environment and natural resources as we know them should be used while at the same time benefitting us. From the researches I carried out in Mozambique as demonstrated in the chapters that constitute this book, it has emerged that [sustainable] conservation is impossible as long as the rural communities who are both users (or at least live near natural resources such as forests, wildlife, fisheries, and most important resource arable land) and caretakers lack knowledge of sustainability and sustainable development. To emphasize this point, in chapters two and seven of this book, I have argued that over-exploitation or poor extraction and mismanagement of 'common properties/resources' have far reaching consequences to the local communities, the nation and also to the global world. While this anticipation of the effects of over-exploitation and mismanagement of the environment and natural resources may seem rather harsh and unsympathetic especially to externally driven community-based natural resource management enthusiasts, it is nevertheless a point that needs to be seriously considered by conservationists/environmentalists the world over. Also, the point is a widely shared perception. In his paper, "Small-scale mining and alluvial gold panning within the Zambezi Basin: An ecological time-bomb and a tinderbox for future conflicts among riparian states", Shoko[235], for example, explores the negative impacts of poor extraction of the resource mineral, gold, in the Zambezi basin, to the local communities, the natural environment and other users along the Zambezi river. In another study, De Georges and Reilly[236] noted with concern that due to poor management of resource, wildlife, during 1999/2000 farm invasions in Zimbabwe, there was a 50% loss in wildlife numbers, a 65% loss of tourism, and a loss of up to 90% of safari hunting on commercial farms[237] from poaching and habitat destruction. As estimated by Unti,[238] by 2007 there was an 80% decline in wildlife in

---

[235] For further discussion on mining in the Zambezi Basin see Shoko, D. S .M. Small-scale mining and alluvial gold panning within the Zambezi Basin: An ecological time-bomb and a tinderbox for future conflicts among riparian states, In Chikowore, G., et al., (Eds). 2002. *Managing common property in an age of globalization-Zimbabwean experiences,* Weaver Press: Harare.

[236] De Georges, A. and Reilly, B. 2007. Politicization of land reform in Zimbabwe: impacts on wildlife, food production and the economy, *International Journal of Environmental Studies,* Vol. 64, No. 5, October 2007, 571–586.

[237] Herbst, W., 2002. Decimation of Zimbabwe's wildlife, 13 June, Wildlife Producers Association. Kubatana.net, The NGO Network Alliance Project – an online community for Zimbabwe activists; Also Available   at: http://www.kubatana.net/html/archive/wild/020613wpa.asp?sector=WI LD; See also *Zimbabwe Independent*, 2002. Safari operators lose 90% of their game,   21   June;   Also   Available   at: http://www.zimbabwesituation.com/june_2002_archive.html.

[238] See Unti, B., 2007, As Zimbabwe's woes mount, Mugabe declares open season on wildlife. The Humane Society of the United States; Available at:

conservancies and game farms, and 60% in national Parks due to mismanagement of resource, wildlife during the farm invasion grip in Zimbabwe. Heath also estimates that there were over 1000 game ranches prior to the radical land reform, in 2006 reduced to about four functional conservancies of 120 properties plus an additional 60 game ranches scattered around the country[239]. There is, therefore, a growing consensus at least in some academic circles that poor extraction and utilization of resources is bad news to natural resource personnel and all concerned parties, especially in developing economies such as Mozambique where these resources are a backbone of the national economies. Yet, even though this is a point that can be easily embraced by many people, conservation has remained a 'hot' issue across the globe.

A glimpse at the literature on conservation in the last two or so decades proves that the subject of conservation has increasingly become topical across the globe. Example of such literature includes, among others, the following books: *Beyond sacred forest – Complicating conservation in Southern Asia (2011), Communities and conservation: Histories and politics of community-based natural resource management (2005), Tropical forest conservation: An economic assessment of the alternatives in Latin America (1998),* and *Conservation biology: The theory and practice of nature conservation, preservation and management (1992).* All these

---

http://www.hsus.org/about_us/humane_society_international_hsi/speci al_programs_projects/mugabe_declares_open_season_on_wildlife.html

[239] See Heath, D. 2006. Game farms before and after radical land reform Zim. Letter, 14 June 2006 to Andre De Georges from Dr Don Heath, former Senior Ecologist (Utilization) Department of National Parks and Wildlife Management (DNPWM), Zimbabwe, currently editor *African Hunter Magazine*, Harare.

texts, in one way or another, show that the advantage of the environment and natural resources over-exploitation enjoyed by the few (those who directly benefit from the exercise) come at a heavy price, especially to the poor who in most cases pay the price. Bearing this in mind, the chapters that make up this volume are in good company in so far as all of them emphasize the need for active participation of all stakeholders in issues concerning environmental and resource conservation and management in Mozambique and beyond.

While the chapters in this book examine different themes and uses different case studies, the fact remains that all of them make a clarion call to sustainable utilization of resources, and to active participation of all actors in the conservation and management of the environment and resources especially the rural communities. Yet while active participation of 'local' members is emphasized, control by the government through responsible ministries remains critical to ensure sustainable utilization of resources. To emphasize this point, different case studies across Mozambique have been employed in this book to demonstrate how lack of government commitment or its reluctance to both empower local communities and provide checks and balances [negatively] impact on the general conservation and management of the environment and resources therein. The point is, however, not emphasized by way of providing an abridged version of the conclusions of the chapter, but by presenting what I believe is the connecting thread that cut across all chapters. To this end, I stressed the point that conservation of the environment and natural resources is all about relational networks and interactions between the state, humans, nonhuman others and the natural environment in general. Since in these relational interactions, power determines who benefits and loses between the state and the governed, I have argued for devolution (by the state to local communities) of some rights to natural resources. These entitlements include, among others, the right to land, right to acquire title deeds and right to benefit directly from the proceeds obtained from land by local communities. Devolution like political economy is best placed to provide answers to the different types of access to natural resources as well as common property.[240]

The connecting thread cutting across conservation discourse

Environment and Natural Conservation has presented different issues and challenges facing environment and natural resource conservation in Mozambique and other developing countries in Africa. I argue that environmental and natural resource conservation can only be done in a sustainable manner if there is at least active participation and recognition of rights, whether political, economic, cultural or social among various stakeholders.

Yet once pointing out the need for the recognition of rights of all stakeholders involved in conservation, what we imply is that environmental and natural resource conservation should be understood in terms of mutual interactions and relationships between the different actors and aspects of conservation. The actors here referred include, among others, the state, humans, and nonhumans (including animals and other such entities).

---

[240] For further discussion on common property and devolution see Kiely, R. and Marfleet, P. (Eds). 1998. *Globalization and the third world*, Routeledge: New York.

Also, in considering interactions and relationships between different actors, there is need to recognize that engaging in sustainable utilization of resources is good in itself, not only at local level but nationally and globally. As such, even when thinking of solutions to the problems affecting conservation of our environment and resources, one should bear in mind that there are no strictly local solutions to the conservation of natural resources and the environment. This 'reality' is premised on the fact that what are sometimes termed local solutions to conservation are indeed international or global solutions and vice versa. To this end, there is always need to dismantle the modernist paradigmatic oppositions such as the global versus the local, and nature versus culture when it comes to issues of environment and natural resource conservation. Symmetrical relations and mutual interactions should be stressed.

Chapters 3 and 4 of this book have discussed in some detail the question of interaction and relationships between humans the natural environment as a whole, especially with other such entities as animals. A critical question that has been raised in chapter 3, for example, is whether nonhumans such as animals have rights just like humans beings or not. This question lies at the heart of the whole debate on environment ethics and resource conservation. Using 'real' examples of animal suffering and destruction of biodiversity and forest resources in particular through veld fires in the face of green revolution program in Mozambique, I have tried to show how biological resources also merit respect even though they can be exploited by humans. It has been demonstrated that uneven power relations between the state, humans and nonhuman others exemplified through poor governance often mediate such unfair exploitations of some actors by others. A similar argument has been pursued in chapter 4, where as a result of gold panning by a minority group, the environment, other resources such as fisheries or aquatic creatures and majority of the people, have and are actually paying a heavy price from pollution and land degradation, among other problems. In fact, besides other humans, both the environment and other natural resources such as fisheries, forestry and wildlife suffer in one way or another in the hands of a minority group involved in the poor extraction of the resource mineral, gold.

The issue of poor resource management, and in particular over-exploitation or poor extraction of resources by the rural communities mainly due to poverty, poor governance, ignorance and lack of resource diversification has been discussed at length in chapter 7. While it has been noted that poverty, lack of resource diversification, low education levels and ignorance contribute to poor conservation of the environment and natural resources, it should be stressed that all these problems zero at governance, in this case, poor governance, where in theory the state says it always acts in the best interests of its people when in practice it does the opposite. To this end, I have cited "political will" on the part of the government, and argued for the need (by the government) to empower rural communities through education, job creation and devolving of 'actual' rights to resources so that the local communities would be equipped with the necessary skills to sustainably exploit the resources at their disposal. Also, a step towards that direction would give the locals a sense of 'real' ownership of resources. This is what has been meant in this volume as active participation of local communities in environmental and resource conservation and management issues.

The issue of local community participation has been discussed in some detail in chapter 5. While acknowledging the need for the government to provide checks and balances in the whole process of environment and resource management, I have emphasized the need for government to devolve rights to resource conservation to the local communities. As highlighted above, devolving rights to local communities have the advantage that it boosts the locals' morale, such that besides feeling a sense of ownership the local community members would recognize themselves as 'real' managers of the resources in their communities.

The final word

This book has explored several interlocking dimensions of environmental and natural resources conservation in Mozambique. To this effect, the book represents an effort, both at theoretical and practical levels, to make conservation of the environment and natural resources sustainable while benefiting local communities, of whom the majority, (about 80 %) of Mozambicans live in abject or extreme poverty, spending less than US$ 1 per day. Yet this effort can never materialize or fully realized as long as asymmetrical relations between the government and the governed persist, and modernistic paradigmatic oppositions such as nature versus culture are cherished. As such, I conclude that the success of conservation of the environment and natural resources in Mozambique, as in many other countries in the region and beyond, largely depends on the strength and capacity of its national institutions to articulate national conservation priorities, interests and goals that are people driven and located within conservation goals at global level.

Bibliography

Abrams, L. J. 1996. 'Review of Status of Implementation Strategy for Statutory Water Committees', *unpublished report*, Department of Water Affairs and Forestry: Pretoria.

Acres, B. D. et al. 1985. 'African dambos: their distribution, characteristics and use', In Thomas, M.F. & Goudie, A.S. (eds.), *Dambos: small channelless valleys in the tropics*, Zeitscrift für Geomorphologie, Supplement band, 52: 63-86.

Africa News Network, 2008. 'Mozambique green revolution will depend on small scale farmers', Mozambique Afrol News, Available online at: *http://www.afrol.com/Mozambique*. (Accessed 27/06/2008).

Afrol.com, 2008. Available online at: *http://www.afrol.com/Mozambique*. (Accessed on August 26, 2011).

Altieri, M. A. 1998. *Agro-Ecology. The Science of Sustainable Agriculture*, Macmillan. New York.

Anstey, S.G. 2001. 'Necessarily vague': The political economy of community conservation in Mozambique', In Hulme, D. and Murphree, M.W. (Eds). *Community conservation in southern and eastern Africa,* James Currey: Oxford.

Anstey, S.G. 2004. 'Governance, natural resources and complex adaptive systems: A CBNRM study of communities and resources in northern Mozambique', In Dzingirai, V. and Breen, C. 2004. *Confronting the crisis in community conservation – Case studies from Southern Africa,* Centre for Environment,

Agriculture and Development, University of KwaZulu-Natal: South Africa.

Anstey, S. G., Abacar, J.A., and Chande, B. 2002. "It's all about power, it's all about money": Governance and resources in northern Mozambique, *Paper presented at the IASCP Conference,* Victoria Falls: Zimbabwe.

Arndt, C. 2006. HIV/AIDS, human capital, and economic growth prospects for Mozambique, *Journal of Policy Modelling* 28 (5) 477–489.

Arundhati, R. 2004. 'How deep shall we dig?' In *Asiatic Society,* Aligarh Muslim University Press, India.

Bambaige, A. 2008. National Adaptation Strategies to Climate Change Impacts: A Case Study of Mozambique, World Human Development Report Office, USA.

Bebbington, A. 1999. Capitals and capabilities: A framework for analysing peasant viability, rural livelihoods and poverty, *World Development,* 27: 2012-44.

Behr, A.L. 1988. *Empirical Research Methods for the Human Sciences* (Second Edition). Durban Butterworths.

Benedikz, K., 2002. Rural Livelihoods at Risk: Land use and coping strategies of war-affected communities in Sri Lanka, *Proceedings on the Conference on International Agricultural Research for Development,* Witzenhausen, Deutscher Tropentag, Oct., 9-11.

Bond, I. 1999. 'Economic incentives for institutional change for the management of natural resources', In Johnson, S. And

Mbizvo, C. (Eds). *Proceedings of the exchange visit workshop for directors*, IUCN-ROSA, Harare.

BMZ, 1992. *Federal Ministry of Economic Co-operation and Development: Ninth Report on German Government Development Policy*, Bonn, German.

Bonner, R. 1993. *At the hand of man*, Alfred Knopf: New York.

Booyson, P. and Tainton, N. M. (eds). 1984. *Ecological effects of fire in South African ecosystems*, Springer-Verlag: Berlin.

Bourdillon, M. 1987. *The Shona peoples: An ethnography of the contemporary Shona with special reference to religion*, 3rd ed, Mambo Press: Zimbabwe.

Boss, J.A. 1999. *Analysing Moral Issues*, Mayfield Publishing Company, Belmont.

Brechin, S.R., P.R. Wilshusen, C.L. Fortwangler and P.C. West. 2003. *Contested Nature*. New York: SUNY Press.

Bullock, A. 1992. Dambo Hydrology in Southern Africa – Review and Reassessment, *Journal of Hydrology*, 134: 373-396.

Burrow, E and Murphree, M. 1998. *Community conservation from concept to practice: A practical framework*. Institute for Development Policy and Management, University of Manchester: UK.

Buscaglia, L. 2010. 'Milk and Honey in Mozambique', Available   at:

http://www.milkandhoney2010.com/201004/mozambique.h
tml (Accessed on Tuesday, April 13, 2010).

Busher, B. 2013. Prosuming conservation, *Paper presented to the Department of Sociology*, University of Cape Town, South Africa.

Cain, J., Moriarty, P., Lynam, T., and Frost, P. 1999. *An update on the integrating modelling strategy, Micro-catchment management and common property resources, 2ⁿᵈ Integrated Modelling workshop, Institute of Environmental Sciences*, University of Zimbabwe and Institute of Hydrology, Harare.

Campbell, M. B. and Luckert, K. M. (Eds). 2002. *Uncovering the hidden harvest- Valuation methods for woodland and forest resources,* Earthscan Publications Limited, London.

Carney, D. 1998. *Sustainable rural livelihoods: What contribution can we make?* Department for International Development, London.

Cavendish, W. 2001. Rural livelihoods and non-timber forest products'. In De Jong, W and Campbell, B. (Eds). *The role of non-timber forest products in socio-economic development,* CABI Publishing: Wallingford.

Cavendish, W. 1999, in Shackleton, C.M. And Shackleton, S.E., 2000. *Direct Use Values Of Secondary Resources Harvested From Communal Savannahs In The Bushbuckridge Lowveld, South Africa. Journal Of Trop. Forest Products 6, 28-47.*

Chambers, R. and Conway, G. 1992. Sustainable Rural Livelihoods: Practical Concepts for the 21ˢᵗ Century. *Discussion Paper 296,* Institute of Development Studies.

CIA, 2009. List of Countries by Unemployment rate, Available @https://www.cia.gov/library/publications/the-world-factbook/fields/2129.html

Chibisa, P. and Rwizi, L. 2009. Traditional Crafts and Rural Livelihoods in Manicaland's Semi-Arid Areas: Implications for Biodiversity and Environment Sustainability, *Journal of Sustainable Development in Africa, Vol 11, Number 2009.*

Child, G. 1995. *Wildlife and people: The Zimbabwean success – How conflict between animals and people became progress for both,* Wisdom Institute: Harare.

Cook, J. and Fig, D. 1995. From colonial to community-based conservation: Environment justice and the national parks of South Africa, *Society in Transition,* 31 (1): 22-36.

Cunningham, A.B., 1993. *Ethics, Ethno botanical Research, And Biodiversity, People And Plants Initiative,* WWF International, Gland.

De Georges, A. and Reilly, B. 2007. Politicization of land reform in Zimbabwe: impacts on wildlife, food production and the economy, *International Journal of Environmental Studies,* Vol. 64, No. 5, October 2007, 571–586.

Doolittle, A. 2007. Native and tenure, conservation, and development in a pseudo democracy: Natural resource conflicts in Sabah, Malaysia, *Journal of Peasant Studies,* 34 (3): 474-97.

Dove, R. M. 1993. The responses of Dayak and Bearded Pig to Mast-Fruiting in Kalimantan: An analysis of nature-culture

189

analogies, In Hladik, M.C. (Ed). *Tropical forests, people, and food*, UNESCO: Paris.

Dove, R. M., Sajise, E. P., and Doolittle, A. A. 2011. *Beyond the sacred forest-Complicating conservation in southeastern Asia*, Duke University Press, London.

Dovie, B.D, Shackleton, C.M., And Witkowski, E.T.F., 2001. Involving local people: Reviewing participatory approaches for inventorying the resource base, harvesting and utilization of non-wood forest products, In *Harvesting of non-wood forest products: Proceedings of FAO/ECE/ILO International Seminar*, Ministry Of Forestry, Turkey Pp. 175-187.

Dzingirai, V. and Breen, C. 2004. 'The community-based natural resource management crisis and research agenda', In Dzingirai, V. and Breen, C. 2004. *Confronting the crisis in community conservation: Case studies from southern Africa*, Centre for Environment, Agriculture and Development, University of KwaZulu-Natal, Harare.

Ellis, F., 1998. Survey article: Household strategies and rural diversification. *Journal of Development Studies*, 35(1): 1-38.

FAO. 1995, *World Agriculture: Towards 2010. An FAO Study*, Ed. N. Alexandratos. FAO, Rome, Italy.

FAO. 1997. Irrigation potential in Africa: A basin approach. *Land and Water Bulletin No. 4*. Rome.

FAO. 2002. FAO program in Mozambique 2002-2006. Discussion paper. Second draft. *Prepared by FAO Representation in Mozambique*, Maputo.

Federal Republic of Nigeria National Policy on Environment (FRNNPE) 2004. *Nigeria's National Policy on Environment*, Nigeria.

Folke, C., T. Hahn, P. Olsson and J. Norberg. 2005. Adaptive governance of social-ecological systems. *Annual Review of Environment and Resources* 30: 441-473.

Geffray, C. 1990. 'La Cause das armes au Mozambique,' Paris : France.

Goudie, A. 1983. *Environmental change,* Clarendon Press: Oxford.

Goudie, A. 1990, *The human impact on the natural environment,* 3rd ed, Basil Blackwell Publishers: UK, pp. 262.

Hanlon, J. 2004. *Do Donors Promote Corruption? The Case of Mozambique.* Third World Quarterly 25 (4), pp. 747-763.

Hatton, J., Couto, M., and Oglethorpe, J. 2001. Biodiversity and war: A case study of Mozambique, *Biodiversity Programme*, WWF, Washington DC, USA.

Health Alliance International Report, 2005. Integrating TB and HIV care in Mozambique: lessons from an HIV clinic in Beira, Beira: Mozambique; Manica district of Mozambique, *2005 Final Report. United Nations Industrial Development Organization*, Vienna, Austria. Available online at: <www.globalmercury.org>.

Herbst, W., 2002. Decimation of Zimbabwe's wildlife, 13June, Wildlife Producers Association: Kubatana.net, The

NGO Network Alliance Project – an online community for Zimbabwe activists. Available online at: http://www.kubatana.net/html/archive/wild/020613wpa.as p?sector=WILD.

Hermele, K. 1988. 'Lands struggles and social differentiation in southern Mozambique: A case study of Chokwe, Limpopo', Uppsala.

Heurtin and Licoppe, Apple Inc. 2001. *Handbook of research on user interface design and evaluation for mobile technology*, Taylor and Harper.

Hobley, M. and Shields, D. 2000. 'The reality of trying to transform structures and processes: forestry in rural livelihoods', *Working Paper 132*, Overseas Development Institute, London.

Human, J. 2000. *Community Forest Management: A Case From India*. Oxfam Publishers: London.

Intergovernmental Panel on Climate Change (IPCC), 2001. Third Assessment Report: Climate Change TAR, http://www.grida.no/publications/ipcc_tar/.

Institute for Poverty, Land and Agrarian Studies (PLAAS). (July 13, 2013). The Distribution of Land in South Africa: An Overview, Article available at PLAAS website: http://www.plaas.org.za/sites/default/files/publications.pdf /No1%20Fact%20check%20web.pdf.

Katerere, Y. 1999. Overview of CBNRM in the region, *Paper presented at the workshop: CBNRM and its contribution to economic*

development in Southern Africa, 3-5 June 1999, Chilo Safari Lodge, Mahenye: Zimbabwe.
Katerere,

Kiely, R. and Marfleet, P. (Eds). 1998. *Globalization and the third world*, Routeledge: New York.

Lele, U. and Adu-Nyako, K. 1991. 'An integrated approach of strategies for poverty alleviation: a paramount priority for Africa', *A paper prepared for the annual meeting of the African Development Bank Group*, Abidjan: Cote d'Ivoire.

Leslie, J. 2008. 'Conservation in Flux: Pursuing Social Resilience in Mozambique and Peru', *Bulletin of the Yale Tropical Resource Institute*, Volume 27.

Little, P. 1999. Environments and Environmentalisms in Anthropological Research: facing the new millennium. *Annual Review of Anthropology*, 29: 253-284.

Lulandala, L. L. 1991. *Agro-forestry potentials for Mozambique*, UNDP/FAO/GOM Project MOZ/88/029, Maputo: Mozambique.

Lundin, I. 2000. *Africa Watch: Will Mozambique Remain a Success Story?* African Security Review Vol 9 No 3.

Mackel, R. 1985. Dambos and related landforms in Africa – an example for the ecological approach to tropical geomorphology. In M.F. Thomas & A. S. Goudie (Eds.), *Dambos: small channelless valleys in the tropics*. Zeitscrift für Geomorphologie, Supplement band, 52: 1-23.

Mackenzie, C. 2006. Forest governance in Zambezia, Mozambique, Mozambique: Chinese Takeaway! *Final Report for FONGZA,* Maputo: Mozambique.

Makamure, D.M. 1970. 'Cattle and Social Status' in Kileff C and Kileff. P (Eds), *Shona Customs*, Mambo Press, Gweru.

Malakata, M. 2007. 'Mozambique aims to lead green revolution', Available at: http://www.scidev.net/en/news/

Manuel, I.R.V. et al. 1999. Exploração artisanal do ouro no distrito de Manica: Degradação ambiental versus desenvolvimento, *Congresso Luso-Mozambique de Engenharia,* Maputo.

Marongwe, N. 2004. 'Traditional authority in community-based natural resource management (CBNRM): The case of Chief Marange in Zimbabwe', In Dzingirai, V and Breen, C. 2004. *Confronting the crisis in community conservation-Case studies from Southern Africa, Centre for Environment,* Agricultre and Development, University of KwaZulu-Natal.

Mawere, M. 2012. Buried and forgotten but not dead': Reflections on 'ubuntu' in environmental conservation in southeastern Zimbabwe, *Afro-Asian Journal of Social Sciences*, Vol. 3, No. 3.2 Quarter II 2012.pp. 1-20.

McNamara, 1995. In World Bank, "Ghana poverty past, present and future", *Report No.14504-GH*, Washington DC: USA.

Medicins Sans Frontiere (MSF)/(Doctors Without borders), 1998. 'Attacks as told by victims MSF'. *MSF Article,* (Retrieved May 10, 2010).

Miller, H.B. 1983. "Platonists" and "Aristotelians" in Miller. H.B. and Williams. W.H (Eds) *Ethics and Animals,* Humana Press, Clifton.

Ministério de Planeamento e Finanças, 2004. *Poverty and Wellbeing in Mozambique: Second National Assessment,* IFPRI, Purdue University.

Mkwanazi, H. 2007. Veld fire Campaign in Lupane. *Environment Africa,* Lupane.

MMSD, 2002. Breaking new ground: mining, minerals and sustainable development, *International Institute for Environment and Development,* London.

Mondlande, D.S. et al. 2002. The Socio-economic impacts of artisanal and small scale mining in the developing countries, *Blackwell Publishers,* Rotterdam, The Netherlands.

Moyana, H. 1984. *Political economy of land in Zimbabwe,* Mambo Press: Gweru.

Moyo, S. 1995. *The land question in Zimbabwe,* Sapes Books: Harare.

Mozambique First National Communication to the United Nations Framework Convention on Climate Change (UNFCCC), 2006, *Submitted to the Secretariat of the Convention in 2006,* Maputo: Mozambique.

Murray, C., 2002. Livelihoods research: Transcending boundaries of time and space. *Journal of Southern African Studies,* 28(3): 489-509.

Murombedzi, J.C. 2003. 'Pre-colonial and colonial conservation practices in Southern Africa and their legacy today', In Whande, W.; Kepe, T. and Murphree, M.W. 2003. *Local communities, enquiry and conservation in Southern Africa*, Africa Resource Trust: Harare.

Murphree, M.W. 1991. Communities as institutions for resource management, *CASS Occasional Paper*, University of Zimbabwe: Harare.

Murphree, M.W. 1995. 'Optimal principles and pragmatic strategies: creating an enabling politico-legal environment for CBNRM', In Rihoy, E. (Ed), *The commons without the tragedy*, SADC Natural Resource Management Conference Report, SADC, Lilongwe.

Natural Resources, Agricultural Development and Food Security (NAF), 2009. International Research Network Report.

Neumann, P. R. 1998. *Imposing wilderness: Struggles over livelihood and nature preservation in Africa*, University of California Press: Berkeley; See also Doolittle, A. 2007. Native and tenure, conservation, and development in a pseudo democracy: Natural resource conflicts in Sabah, Malaysia, *Journal of Peasant Studies*, 34 (3): 474-97.

Newitt, M. 1995. *A history of Mozambique*, Hurst and Company, London.

Nhantumbo, I., Chonguisa, E., and Anstey, S. 2003. Community-based natural resource management in

Mozambique: The challenges of sustainability, *Report to IUCN-SASUSG*, Harare.

Nhantumbo, I., Dent, J.B. and Kowero, G.S. 2001. Goal Programming: application in the management of the Miombo Woodlands in Mozambique, *European Journal of Operational Research* 1332 (2): 310-322.

Nhantumbo, I., Foloma, M., Puna, N. 2004. Comunidades e Maneio dos Recursos Naturais: Memórias da III Conferência Nacional Sobre o Maneio Comunitário dos Recursos Naturais, Maputo, Mozambique, 21 – 23 de Julho de 2004, UICN, vol. I.

Nigerian Institute of Social and Economic Research (NISER). 2009. *Poverty Alleviation in Nigeria*, NISER, Ibadan.

Nkhata, A. 2004. 'Devolution and natural resources management in Zambia: Transforming rural communities into gatekeepers without authority', In Dzingirai, V. and Breen, C. (Eds). 2004. *Confronting the crisis in community conservation- Case studies from southern Africa,* Centre for Environment, Agriculture and Development, University of KwaZulu-Natal.

O'Connor, D.J. 1985. (Ed). *A Critical History of Western Africa,* Free Press, New York.

OECD, 2009. In Makochekanwa, A. and Kwaramba, M. 2009. *State Fragility: Zimbabwe's horrific journey in the new millennium.* A Research Paper Presented at the European Report on Development's (ERD), Accra: Ghana.

Ogwumike, F. O. 1991. A Basic Needs Oriented Approach to the Measurement of Poverty in Nigeria, *NJESS*, Vol. 33, no 2, 105-119.

PROAGRI, 1997. *The forestry and wildlife sector,* MINADER, Maputo.

Raghavan, S.M. 1999. Animal Liberation and "Ahims" in Boss. J.A.(Ed) *Analysing Moral Issues,* Mayfield Publishing Company, Belmont.

Rakodi, C. and Loyd-Jones, T., 2002. *Urban Livelihoods: A People-Centred Approach to Reducing Poverty.* Earthscan Publications Ltd: London.

Rambe, P. and Mawere, M. 2011. Barriers and constraints to epistemological access to online learning in Mozambique Schools, *International Journal of Politics and Good Governance,* 2 (2.3 Quarter III): 1-26.

Rappaport, R.A. 1979. *Ecology, meaning, and religion,* Berkeley, CA: North Atlantic.

Rockström, J. 2000. Water resources Management in smallholder Farms in Eastern and Southern Africa: An Overview, *Phys. Chem. Earth (B)*, Vol 25, No. 3, pp275-283.

Rodman, S. and Gatu, K. 2008. 'A Green Revolution in Southern Niassa, Mozambique?: A field Study from a small Farmer Perspective about Possibilities and Obstacles for a Green Revolution', Paper presented at Växjö Universitet.

Roesch, O. 1989. 'Economic reform in Mozambique: notes on stabilization, war, and class formation' *Paper presented in Taamuli*, Dar es Salaam.

Rural Poverty Report-Mozambique, 2007, *Censo*, Mozambique.

SAFIRE, 2003. Annual Report 2002, Harare. SAFIRE. Available at : http://www.safireweb.org).

Salomão, A. 2006. 'Towards people-centred woodland management in Mozambique: can this make a difference?: community-based natural resources management in miombo forest in Mozambique and the fight against poverty', *Draft Paper*, Maputo, Mozambique.

Salamao, A. 2002. Participatory natural resources management in Mozambique: An assessment of legal and institutional arrangements for community- based natural resources management, *World Resources Institute*, Washington D.C.

Schubert, R. 1994. Poverty in developing countries: Its definition, extent, and implications, *ECONOMICS* FRG, Vol. 49/50.

Scoones, I., 1998. Sustainable rural livelihoods, a framework for analysis, *Institute of Development Studies*, Working Paper 72.

Sen, A. and Jean, D. K. 1989. *Hunger and public action*. Clarendon Press, Oxford.

Shackleton, C.M. and Shackleton, S.E., 2000. Direct use values of secondary resources harvested from communal savannahs in the Bushbuckridge lowveld, South Africa, *Journal Of Trop. Forest Products 6, 28-47.*

Shackleton, C.M., Shackleton, S.E. and Cousins, B. 2001. The role of land-based strategies in rural livelihoods: The contribution of arable production, animal husbandry and natural resource harvesting in communal areas in South Africa, Development Southern Africa, 18 (5): pp. 581-604.

Shandro, J.A. et al., 2009. Reducing mercury pollution from artisanal gold mining in Munhena, Mozambique, *Journal of Cleaner Production,* Vol 17 (5) 525-532.

Shoko, D.S.M. Small-scale mining and alluvial gold panning within the Zambezi Basin: An ecological time-bomb and a tinderbox for future conflicts among riparian states, In Chikowore, G., et al., (Eds). 2002. *Managing common property in an age of globalization-Zimbabwean experiences,* Weaver Press: Harare.

Singer, P. 1993. *Practical Ethics,* Cambridge: Cambridge University Press.

Southern African Development Community (SADC), 1996. Short term consultancy in agro-forestry diagnosis and design for some SADC countries, Vol. 2, Botswana, 4 Mozambique, and 6 South Africa, *SADC Energy Sector TAU*, Luanda.

Soto, B. 2003. Protected areas management in Mozambique, Report for IUCN-SASUSG, Harare.

Spector, B, Schloss M, Green S, Hart E and T Ferrell, 2005. *Corruption Assessment: Mozambique.* USAID.

Spence, C.F. 1963. *Mozambique: East African province of Portugal,* Howard Tommins, Cape Town.

Spiegel, S.J. et al., 2006. Mercury reduction in Munhena, Mozambique: homemade solutions and the social context for change, International Journal of Occupational and Environmental Health, *Scopus,* 15, p. 215–221.

Swain, E. B. et al., 2007. Socio-economic consequences of mercury use and pollution, *Ambio* 36, pp. 46–61.

Tanner, C. 1993. 'Land disputes and ecological degradation in an irrigation scheme: A case study of state farm divestiture in Chokwe, Mozambique', *Paper presented to Land Tenure Centre:* University of Wisconsin-Madison.

Tascconi, L. Tisdell, C. 1993. Holistic Sustainable Development: Implications for Planning Processes, Foreign Aid and Support for Research. *Third World Planning Review,* 14: 4.

Telmer, K. and Veiga, M. M. 2008. World emissions of mercury from artisanal and small scale gold mining. In: Pirrone, N. and Mason, R. (Eds). *Mercury fate and transport in the global atmosphere: measurements, models and policy implications.* Interim Report of the UNEP Global Mercury Partnership on Mercury Air Transport and Fate Research.

Tello, 1996. In Anstey, S.G. 2004. 'Governance, natural resources and complex adaptive systems: A CBNRM study of communities and resources in northern Mozambique', In Dzingirai, V. and Breen, C. 2004. *Confronting the crisis in community conservation – Case studies from Southern Africa,* Centre for Environment, Agriculture and Development, University of KwaZulu-Natal.

Tembe, M. J. 2008. 'Indigenous vegetables and legumes' importance, utilization and marketing in Gaza Province, Mozambique', Paper presented at the First Scientific and Technological Journeys of Mozambique. *Lutheran World Federation, Mozambique:* Maputo. Available online at: http://www.mct.gov.mz/pls/portal/url/ITEM/5017142A15 9E9D4CE040007F01004B9.

UNCED. 1992. United Nations Conference on Environment and Development, Final Advanced Version of Agenda 21, Chapter 11, *Combating Deforestation,* Washington DC, United Nations publishers.

UNICEF Report, 2011. "Impact of environmental degradation and emergencies on children in Mozambique-Part 2", Available @ *http://www.unicef.org/mozambique/index.html.*

United Nations Environment Program (UNEP), 1999. GEO-2000 global environmental outlook. Nairobi.

United Nations Environment Program (UNEP), 2002. Vital climate graphics Africa (available at www.grida.no).

United Nations Development Program (UNDP) Human Development Report, 2009. Mozambique, http://hdrstats.undp.org/en/countries/data_sheets/cty_ds_MOZ.html.

United Nations Report, 2008. 'The food and Agricultural Organization of the United Nations', The Nutrition Country Profile. Werichannel news http://werichannel.wordpress.com/2008/09/18_scores_die-inmozambican-_veld-fires/

Unti, B., 2007. 'As Zimbabwe's woes mount, Mugabe declares open season on wildlife: The Humane Society of the United States', Available at: http://www.hsus.org/about_us/humane_society_internation al_hsi/special_programs_projects/mugabe_declares_open_se ason_on_wildlife.html.

Veiga, M.M. and Baker, R. 2004. Protocols for environmental and health assessment of mercury released by artisanal and small scale miners, *Report to the Global mercury project: removal of barriers to introduction of cleaner artisanal gold mining and extraction technologies*, GEF/UNDP/UNIDO.

Vogl, R. J. 1974. Effects of fires on grasslands, In Kozlowski, T. T. and Ahlgren, C. C. (eds), Fire and ecosystems, Academic Press: New York, pp. 139-94.

Von der Heyden, C. J. 2004. The hydrology and hydrogeology of dambos: a review, *Progress in Physical Geography*, 28: 544-564.

Watson, J.P. 1964. A soil catena on granite in southern Rhodesia, I. Field observations. *Journal of Soil Science*, 15: 238-250.

Wignaraja, P.A., Hussain, A., Sethi, H., and Wignaraja, G., (eds). 1991. *Participatory Development,* Oxford University Press: Oxford.

Williams, J. R. et al., 2008. *Financial and managerial accounting,* McGraw-Hill: Irwin.

Wirbelaeur, C., Mosimane, A.W., Mabumnda, R., Makota, C., Khumalo, A., and Nanchengwa, M. 2005. A Preliminary Assessment of the Natural Resource Management Capacity of Community Based Organizations in Southern Africa Cases from Botswana, Mozambique, Namibia, Zambia and Zimbabwe, *The Regional CBNRM Project,* WWF Southern Africa Regional Programme Office, Harare, Zimbabwe.

Wisner, B. 1977. Agriculture *in Mozambique. Science for People* 34, London.

Wolmer, W. 2007. *From Wilderness Vision to Farm Invasions: Conservation and Development in Zimbabwe's South-East Lowveld,* Weaver Press, Harare.

World Bank, 2004. *World Development Report: Making Services Work for Poor People.* Washington DC: World Bank.

World Bank, 2009. *Development and Climate Change,* The World Bank group at work. Washington DC 20433.

World Bank. 2010. Extreme poverty rates continue to fall, Available online at: http://www.worldbank.org. (Accessed on September 6, 2011).

World Commission on Environment and Development, 1987. *Our common future,* Oxford University Press: Oxford.

Yi-fu Tuan. 1971. Man and nature, *Commission on College Geography Resource Paper, 10.*

Zacarias, R. & Manuel, I. 2003. Assessment of mercury use in artisanal gold mining in the mining in the Manica district of Mozambique, In: Artisanal and small-scale mining in developing countries, *Urban health and development bulletin,* Vol. 6 (4) 57-61.

www.ingramcontent.com/pod-product-compliance
Lightning Source LLC
Chambersburg PA
CBHW060036030426
42334CB00019B/2359